DEVELOPMENTS IN RUBBER TECHNOLOGY—4

CONTENTS OF VOLUMES 2 and 3

Volume 2: Synthetic Rubbers

1. Trends in the Usage of Rubbery Materials. J. A. BRYDSON

2. Styrene–Butadiene Rubber. J. A. BRYDSON

3. Developments in Acrylonitrile–Butadiene Rubber (NBR) and Future Prospects. H. H. BERTRAM

4. Ethylene–Propylene Rubbers. L. CORBELLI

5. Developments with Polychloroprene. J. C. BAMENT and J. G. PILLOW

6. Butyl and Halogenated Butyl Rubbers. W. D. GUNTER

7. Silicone Rubbers. R. J. CUSH and H. W. WINNAN

8. Synthetic Polyisoprene Rubbers. M. J. SHUTTLEWORTH and A. A. WATSON

   Index

Volume 3: Thermoplastic Rubbers

1. Thermoplastic Rubbers—An Introductory Review. J. A. BRYDSON

2. Ethylene–Propylene Rubber–Polypropylene Blends. R. RANALLI

3. Styrenic Block Copolymers. F. MISTRALI and A. PRONI

4. Ethylene–Vinyl Acetate Copolymers. G. W. GILBY

5. Thermoplastic Polyurethanes. D. J. HARROP

6. Polyester Elastomers. I. R. LLOYD

7. Natural Rubber Systems. D. J. ELLIOTT

   Index

DEVELOPMENTS IN RUBBER TECHNOLOGY—4

Edited by

A. WHELAN and K. S. LEE

London School of Polymer Technology
(formerly National College of Rubber Technology),
Holloway, London, UK

ELSEVIER APPLIED SCIENCE
LONDON and NEW YORK

ELSEVIER APPLIED SCIENCE PUBLISHERS LTD
Crown House, Linton Road, Barking, Essex IG11 8JU, England

Sole Distributor in the USA and Canada
ELSEVIER SCIENCE PUBLISHING CO., INC.
52 Vanderbilt Avenue, New York, NY 10017, USA

WITH 91 TABLES AND 113 ILLUSTRATIONS

© ELSEVIER APPLIED SCIENCE PUBLISHERS LTD 1987
Softcover reprint of the hardcover 1st edition 1987
British Library Cataloguing in Publication Data

Developments in rubber technology. — 4
1. Elastomers   2. Rubber
678   TS1925

The Library of Congress has cataloged this work as follows:

Developments in rubber technology. — 1- — London: Applied
Science Publishers, 1979–

   v.: ill.; 23 cm. — (Developments series)
ISSN 0262-1584 = Developments in rubber technology.

   1. Rubber—Collected works.   2. Rubber, Artificial—Collected works.   I.
Series.
[DNAL: 1. Rubber industry and trade—Periodicals]
TS1870.D49   678'.2'05—dc19   84-644558
ISBN-13: 978-94-010-8037-8   e-ISBN-13: 978-94-009-3435-1
DOI: 10.1007/978-94-009-3435-1

The selection and presentation of material and the opinions expressed are the sole
responsibility of the author(s) concerned.

Special regulations for readers in the USA
This publication has been registered with the Copyright Clearance Center Inc. (CCC),
Salem, Massachusetts. Information can be obtained from the CCC about conditions
under which photocopies of parts of this publication may be made in the USA. All other
copyright questions, including photocopying outside of the USA, should be referred to
the publisher.

PREFACE

This volume, the fourth in a series which began in 1979, covers a greater variety of subjects than any previous single volume. The basis of selection has been topical interest; hence the tailor-making of polymers to develop specific properties, methods of improving compound processability and the use of rubbers in the oil industry are featured alongside a discussion of safety aspects.

We have again sought the cooperation of the foremost authorities on the chosen subjects and have been delighted at the response which has yielded a list of authors of international repute.

A.W.

K.S.L.

CONTENTS

Preface v

List of Contributors ix

1. Recent Developments in Synthetic Rubbers by Anionic
   Polymerization 1
   I. G. HARGIS, R. A. LIVIGNI and S. L. AGGARWAL

2. Advances in Nitrile Rubber (NBR) 57
   P. W. MILNER

3. Epoxidized Natural Rubber 87
   C. S. L. BAKER and I. R. GELLING

4. Process Aids and Plasticizers 119
   B. G. CROWTHER

5. A Review of Elastomers Used for Oilfield Sealing
   Environments 159
   W. N. K. REVOLTA and G. C. SWEET

6. Using Modern Mill Room Equipment 193
   H. ELLWOOD

7. Quality Requirements and Rubber Mixing 221
   P. S. JOHNSON

8. Health and Safety 253
   B. G. WILLOUGHBY

Index 307

LIST OF CONTRIBUTORS

S. L. AGGARWAL

GenCorp, Research Division, 2990 Gilchrist Road, Akron, Ohio 44305, USA

C. S. L. BAKER

Malaysian Rubber Producers' Research Association, Tun Abdul Razak Laboratory, Brickendonbury, Hertford SG13 8NL, UK

B. G. CROWTHER

Malaysian Rubber Producers' Research Association, Tun Abdul Razak Laboratory, Brickendonbury, Hertford SG13 8NL, UK

H. ELLWOOD

Farrel Bridge Ltd, Queensway, Castleton, PO Box 27, Rochdale, Lancashire OL11 2PF, UK

I. R. GELLING

Malaysian Rubber Producers' Research Association, Tun Abdul Razak Laboratory, Brickendonbury, Hertford SG13 8NL, UK

I. G. HARGIS

GenCorp, Research Division, 2990 Gilchrist Road, Akron, Ohio 44305, USA

P. S. JOHNSON

   Polysar Ltd, Sarnia, Ontario N7V 7U1, Canada

R. A. LIVIGNI

   GenCorp, Research Division, 2990 Gilchrist Road, Akron, Ohio 44305, USA

P. W. MILNER

   Compagnie Française Goodyear, Avenue des Tropiques, ZA Courtaboeuf, BP 31, 91941 Les Ulis Cedex, France

W. N. K. REVOLTA

   Du Pont (UK) Ltd, Maylands Avenue, Hemel Hempstead, Herts HP2 7DP, UK

G. C. SWEET

   Du Pont (UK) Ltd, Maylands Avenue, Hemel Hempstead, Herts HP2 7DP, UK

B. G. WILLOUGHBY

   Rapra Technology Ltd, Shawbury, Shrewsbury, Shropshire SY4 4NR, UK

Chapter 1

RECENT DEVELOPMENTS IN SYNTHETIC RUBBERS BY ANIONIC POLYMERIZATION

I. G. Hargis, R. A. Livigni and S. L. Aggarwal

GenCorp, Research Division, Akron, Ohio, USA

1. INTRODUCTION

Although styrene–butadiene rubbers made in organic solvents (commonly called solution SBR) were introduced commercially in early 1960, their use has been limited to only a few structural varieties. Emulsion SBRs in blends with *cis*-BR (high *cis*-polybutadiene rubbers) were capable of meeting several performance requirements until the demand for low rolling resistance of tyre treads came to prominence in the 1970s. Then the potential of solution rubbers, with their much greater control of molecular structure, surfaced to meet the new requirements for improvements in rolling resistance and wet traction.

The objective of this chapter is to provide a survey of the technological developments in solution rubbers prepared by homogeneous anionic polymerization. Emphasis will be on butadiene and styrene monomers, and catalyst systems based on organolithiums and their complexes with Lewis bases plus some unpublished work on alkaline earths from the authors' laboratory. The rubbers thus obtained offer means for control of molecular structural features of polybutadiene and poly(butadiene-co-styrene) that directly relate to the functional properties of tyre rubbers.

1

2. POLYMER SYNTHESIS AND CHARACTERIZATION

2.1. Early Work on Diene Polymerizations

Early work bearing on organolithium-initiated diene polymerization was reported by Ziegler and co-workers in 1934.[1-3] It was demonstrated that the organolithium adds to the diene double bond to give an intermediate capable of reacting further with the diene.

The characteristics of the organolithium-initiated polymerization of the dienes was clearly established by Morton and co-workers. It was demonstrated that the polymerizations are of the 'living' type, which take place without a chain termination reaction in the absence of impurities.[4,5] The implications of this behaviour in preparing polymers of controlled structure are well recognized today.

Homopolymers and copolymers with tailored-molecular structure, prepared using organolithium, have recently generated much interest because of their potential for improving rolling efficiency, wet traction and durability of tyre treads.

2.2. Organolithium Initiation of Dienes

Anionic polymerization requires unsaturated monomers having substituents that stabilize the negative charge on the active centre (electrophilic substituents) and initiators based on the most electropositive elements. Among initiators for diene polymerizations, organolithiums are nearly ideal compounds. This stems from some of their main characteristics listed in Table 1. Solubility in non-polar solvents results from their covalent character and their ability to form electron-deficient bonding giving rise to associated forms, usually tetramers or hexamers.

TABLE 1

CHARACTERISTICS OF ORGANOLITHIUM POLYMERIZATION CATALYSTS

---

(1) Solubility in non-polar solvents.
(2) Associated structures.
(3) Reactivity varies with Lewis bases.
(4) High 1,4-polydienes.
(5) Living Polymerization.
(6) Control of molecular weight and molecular weight distribution.
(7) Versatility in solution polymerization.

TABLE 2
STRUCTURAL CONTROL IN SOLUTION STYRENE–BUTADIENE RUBBERS USING ANIONIC CATALYST SYSTEMS

(1) Composition.
(2) Monomer sequence distribution (random, tapered, block).
(3) Microstructure variation in polybutadiene microstructure (vinyl, *cis* and *trans*).
(4) Molecular weight and molecular weight distribution.
(5) Molecular chain architecture (linear or star).
(6) Addition of functional groups.

Lewis bases coordinate with the Li^+ ion, breaking up the aggregate. In general, complexing with Lewis bases results in increasing the reactivity of organolithiums and greatly alters, as discussed later in this chapter, the polydiene structure.

The most common Lewis bases used in organolithium-initiated polymerizations are aprotic amines and ethers. More strongly complexing agents include hexamethylphosphoramide, polyglycol dimethyl ethers, cyclic ethers (crown ethers),[6] and dipiperidinoethane (DIPIP).[7]

An important feature of anionic polymerizations with organolithiums is that they can take place in the complete absence of a termination reaction. Organolithiums are stable species in hydrocarbon solvents at normal polymerization temperatures (20–50°C). This is the reason the polymers are commonly referred to as 'living polymers'. This special characteristic coupled with the association features of the propagating centre are mainly responsible for the high degree of structural control for diene rubbers prepared using butyllithium, as summarized in Table 2. In addition to control of polymer composition, control of molecular weight, molecular weight distribution, diene microstructure (geometric isomerism of diene units in the polymer chain), monomer sequence distribution, molecular chain architecture and chain end derivatization can conveniently be achieved. Thus, the use of alkyllithiums as polymerization catalysts offers a highly versatile method for molecular engineering of rubbers of controlled structures.

2.3. Microstructure Variations
Having two conjugated double bonds, butadiene can result in various isomeric structures in the polybutadiene chain, as shown in Fig. 1. When both double bonds participate in 1,4-addition, the remaining double bond in the polymer backbone can exist as two geometric

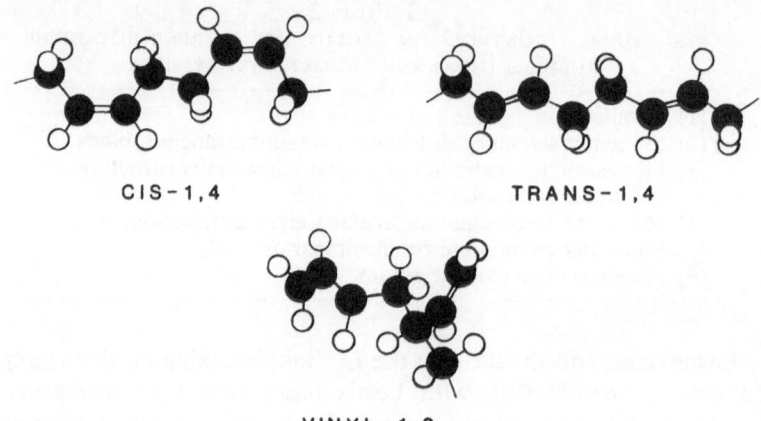

FIG. 1. Polybutadiene isomers.

isomers, namely *cis*-1,4- and *trans*-1,4-units. Addition across only one of the double bonds produces the vinyl or 1,2-structure. Three stereoisomers of 1,2-polybutadiene are possible: isotactic, syndiotactic and heterotactic (atactic). As far as is known, the homogeneous anionic polymerization of butadiene appears to have no effect on the stereoisomerism of the asymmetric carbon in the vinyl structure, in that only the atactic structure has been reported.[8]

Values of glass transition temperature (T_g) and crystalline melting temperature (T_m) for the various polybutadiene structures are given in Table 3. Crystallinity arising from chain stucture regularity can occur

TABLE 3

GLASS TRANSITION TEMPERATURE AND MELTING TEMPERATURE OF POLYBUTADIENES

Polybutadiene	T_m (°C)	T_g (°C)
cis-1,4	2	−113
trans-1,4	—	−88
Form I	55	—
Form II	145	—
Atactic-1,2	—	−4
Syndiotactic-1,2	156	—
Isotactic-1,2	128	—

in all of these forms. Only the atactic vinyl and *cis*-1,4-polybutadiene (*cis*-BR) are amorphous at room temperature in the relaxed state. It is possible, of course, to decrease crystallinity in polybutadienes by introducing irregularity in a random manner, for example by introducing mixed microstructures and/or styrene units that can interrupt crystallizable sequence lengths. Strain-induced crystallization may be achieved in polybutadienes with crystallizable sequences of suitable length under special conditions (Section 3.5.2).

The choice of solvent and/or polar modifier used for organolithium polymerizations greatly affects the chain microstructure of polydienes, comonomer sequence length distribution in styrene–butadiene copolymerizations, and the rate of polymerization. Generally, those modifiers (without bulky substituents) which make it possible to increase the rate of polymerization also result in a higher vinyl content and significantly alter the reactivity ratios for styrene and butadiene. In many circumstances, a more random and less 'blocky' copolymer is produced. On a molecular level, these effects are related to greater charge separation of the polybutadienyl or polystyryl carbanion and the Li^+ cation.

The influence of polar modifier on the microstructure of polybutadiene prepared with butyllithium (BuLi) is summarized in Table 4. It

TABLE 4

EFFECT OF VARIOUS MODIFIERS ON VINYL CONTENT OF POLYBUTADIENES

Modifier	Molar ratio Modifier/BuLi	Microstructure (%)[a]		
		Vinyl	trans	cis
None[b]	—	8	57	35
Triethylamine[c]	270	37	39	24
Diethyl ether[b]	6	21	50	29
THF[b]	1	35	40	25
TMEDA[c]	1	76	16	8
DIPIP[d]	1	99	1	—

[a] Prepared in *n*-hexane at 20–30°C at $[Bd]_0 = 0.5M$ and $[BuLi]_0 = 2 \times 10^{-4} M$.
[b] Morton, M., *Anionic Polymerization: Principles and Practice*, Academic Press, New York, 1983.
[c] Ref. 11.
[d] Ref. 7.

can be seen that the bidentate polar ligands, tetramethylethylenedi-amine (TMEDA) and DIPIP, are capable of producing polymers with high vinyl unsaturation. A vinyl content of almost 100% is obtained in the presence of DIPIP.[9] Uraneck[10] and Antkowiak[11] have reported that in the presence of these polar modifiers lower polymerization temperatures favour higher amounts of vinyl structure. However, in the absence of polar modifiers the diene structure is only somewhat dependent on temperature. For the bidentate ligands, the magnitude of the temperature influence is greater than for the monodentate ligands.[11]

Polydiene microstructure is also dependent on the ratio of con-centrations of monomer to organolithium. This is true for both non-polar and polar media. As catalyst levels are reduced in a non-polar system, cis-1,4-values as high as 86% have been observed in high molecular weight polybutadienes prepared in the absence of any solvents. This compares with 96% cis-1,4-content in similarly prepared polyisoprenes. In the case of high-vinyl BRs (HVBRs) prepared with a polar modifier, such as DIPIP, a decrease in alkyllithium (RLi) concentration at constant ratio of DIPIP/RLi produces smaller amounts of vinyl structure.[12]

In addition to the normal 1,4- and 1,2-structures of polybutadiene, the use of certain polar modifiers, e.g. TMEDA and tetrahydrofuran (THF), can result in the formation of vinylcyclopentane units (Fig. 2).[13–16] This cyclic structure can be as high as 45 wt% in VBRs prepared by the addition of butadiene at a rate such that intramolecu-lar cyclization can compete effectively with chain propagation. No appreciable amounts of cyclic structure are found in VBRs prepared by batch polymerization.[14] How this cyclic structure influences the dynamic mechanical properties needs to be studied.

As discussed above, the amount of 1,2-addition in polybutadiene rubbers (BRs) can be varied from 8 up to 100% depending on the type of modifier and conditions used in organolithium polymerizations. This process versatility provides an important way to control T_g of BR and SBR rubbers. Figure 3 shows the relationship between increasing vinyl

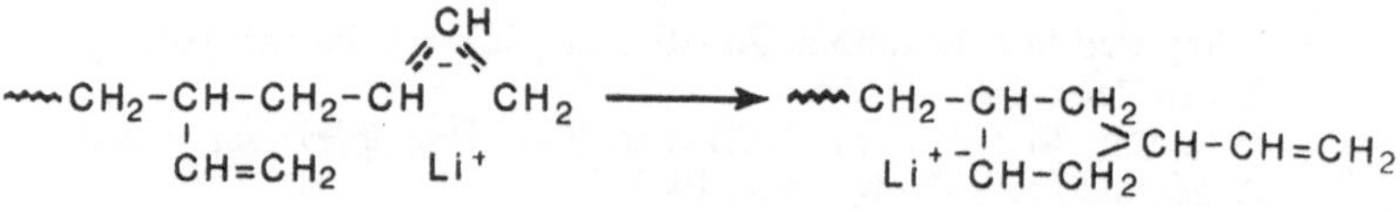

FIG. 2. Proposed mechanism of vinylcyclopentane formation in vinyl-BRs.

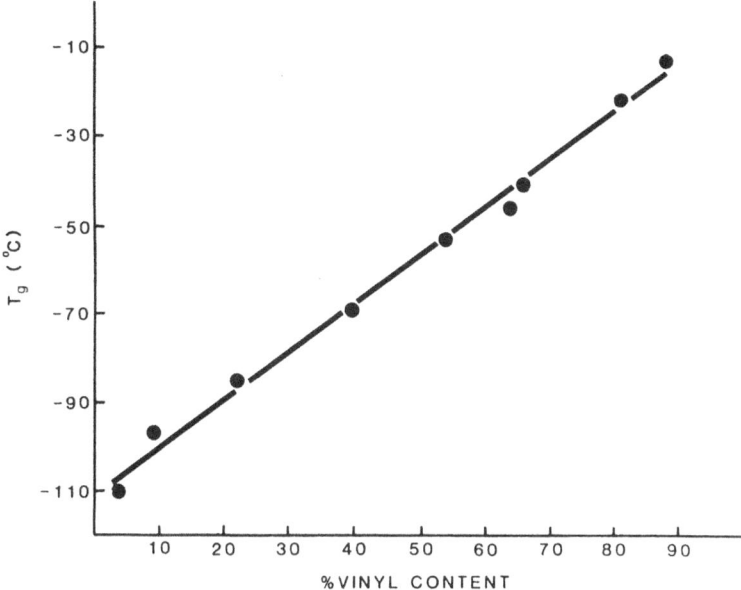

FIG. 3. Glass transition versus percentage vinyl content of polybutadienes.

content of BR (uncrosslinked) and T_g. As expected, T_g increases approximately linearly with an increase in the vinyl content of the polymer. In terms of the functional properties of tyre tread rubbers (discussed in Section 4), the T_g of the rubber can be considered an appropriate material property related to both wet traction and the abrasion resistance of a rubber.[17]

2.4. Control of Comonomer Sequence Distribution

Although the different isomers present in polybutadiene prepared by using alkyllithium catalysts have random placements, as shown by carbon-13 nuclear magnetic resonance (^{13}C NMR),[18] solution SBRs have been produced with four basic types of sequence length of the comonomer units, as shown schematically in Fig. 4.

As previously mentioned, we are concerned in this chapter with essentially random copolymers of styrene and butadiene produced by solution polymerization. In terms of the distribution of styrene and butadiene units, random solution copolymers are considered as having only small variations in the sequence length distribution of the

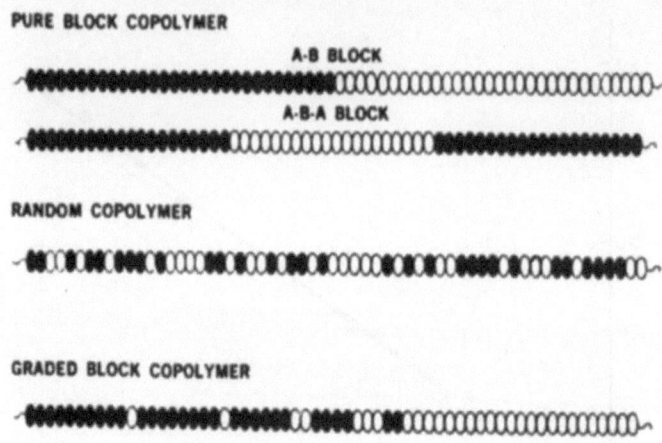

FIG. 4. Schematic representation of copolymer structures.

monomer units along the chain, and between different polymer molecules in a sample.[19] This compares with uniform random copolymers in which the monomer distribution closely follows that expected from random statistical calculations and does not vary inter- and intra-molecularly—a situation that is approached in SBRs prepared by conventional free-radical emulsion polymerization (E-SBRs).

Several authors have reported that when butadiene and styrene are polymerized together, with an organolithium initiator in hydrocarbon solvent, there is a great preference for butadiene and relatively little styrene monomer is polymerized until butadiene is almost consumed.[20–22] This occurs even though the styrene monomer exhibits the faster rate of homopolymerization. At the latter stage of the copolymerization when the supply of butadiene nears depletion, then styrene begins to become incorporated into the chain. The resulting chain structure is referred to as a graded (or tapered) block copolymer having a block rich in butadiene.

Special polymerization approaches are necessary to prepare random solution SBRs with organolithium in hydrocarbon solvent. The main polymerization methods for producing random copolymers, or containing only short (non-blocky) styrene sequences, are given in Table 5. For convenience, the procedures are divided into those that control the polymerization process in order to maintain a high styrene/butadiene (Bd) ratio in the hydrocarbon media, and those that

TABLE 5

APPROACHES FOR PREPARATION OF RANDOM SOLUTION
STYRENE–BUTADIENE RUBBERS

Control of monomer feed composition	Control of reactivity ratios
Continuous Bd addition.	Polar organic modifiers (ethers and amines).
Continuous Bd and styrene addition.	Metal alkoxides
	High polymerization temperature.

use randomizers (polar modifiers) or very high polymerization temperatures to increase the ionic character of the carbon–Li bond. In this way, the reactivity ratios of butadiene and styrene become more favourable for copolymerization. Control of comonomer feed composition and the use of very high polymerization temperatures are typically used to produce SBRs having vinyl contents of about 10%, whereas the use of certain ethers and amines raises the vinyl content and, consequently, T_g. Ionic-type modifiers such as potassium tert-butoxide have been reported to produce random SBRs, at normal polymerization temperatures (20–50°C), having comparably low vinyl contents.[23]

Some of the recent work in the authors' laboratory has shown that catalyst compositions of barium alkoxide salts combined with BuLi as well as with complexes of Mg–Al alkyls also result in low-vinyl solution SBRs (LVSBRs) in hydrocarbon solvent.[24] Such elastomers have a butadiene portion of trans-1,4-content as high as 90% and a vinyl content as low as 2%. The comonomer placement in the copolymers is predominantly random. The development and properties of these rubbers will be described more fully in Sections 3 and 4, respectively.

2.5. Characterization of Styrene Sequence Distribution in Solution Styrene–Butadiene Rubbers by Anionic Polymerization

Before the advent of NMR, qualitative information related to monomer sequence distribution was obtained primarily by the chemical method of Kolthoff.[25] In this procedure, the copolymer is oxidatively degraded at the double bonds of the polybutadiene segments. Polystyrene blocks are not attacked and are isolated and determined gravimetrically as block styrene. Unfortunately, sequencing information pertaining to short segments of styrene are not obtainable by this method.

Nuclear magnetic resonance has proved to be an excellent method that provides detailed sequence data.[26-28] Although only qualitative information on styrene sequence distribution was first possible from proton NMR spectra, recent advances in NMR techniques and spectrometers produce high-resolution spectra capable of providing quantitative information on sequence lengths of styrene in SBR. For example, using a 200 MHz superconducting NMR spectrometer, it is possible to distinguish styrene sequence lengths consisting of isolated units, styrene diads, styrene triads, short blocks of styrene (4–7 consecutive units) and long blocks of styrene (≥8 consecutive units). Proton NMR spectra (Fig. 5) of the styrene ring protons in two low vinyl (10%) solution SBRs (denoted as (A) and (B) in the figure) show the various types of styrene units present, indicated at I = isolated, D = diads, and polystyrene. It is interesting to note that NMR can

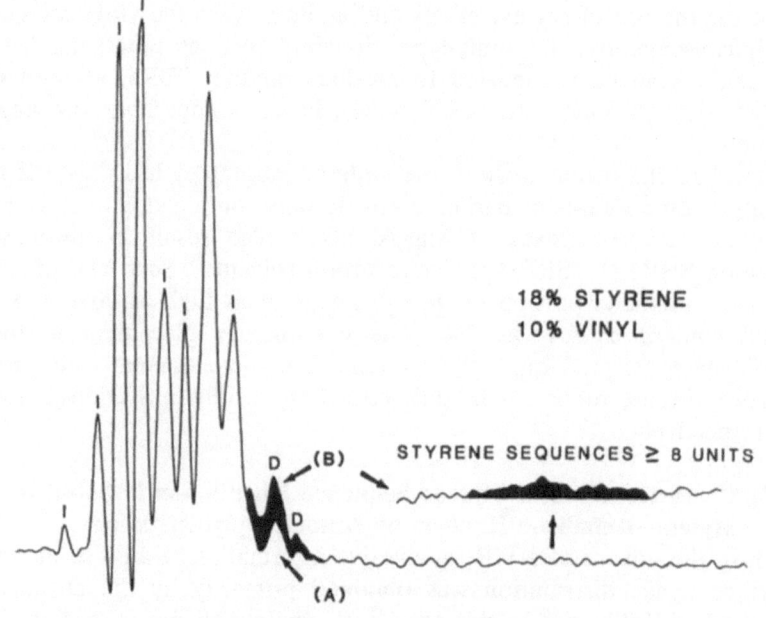

FIG. 5. Proton NMR spectra of two solution SBRs: polymer (A) more random than polymer (B). I, isolated styrene units; D, styrene diads.

TABLE 6
EFFECT OF STYRENE CONTENT AND MOLE RATIO OF TMEDA/BuLi ON THE
AMOUNT OF STYRENE INCORPORATED AS LONG SEQUENCES IN SBR

Sytrene (wt%)	TMEDA/BuLi	Vinyl (%)	Long Sequences of styrene	
			Polymer (wt%)	Total styrene (wt%)
20	0·6	45	0	0
25	0·6	45	1·0	4
30	0·6	45	2·1	7
50	0·6	45	19·5	39
50	1·2	60	0	0

distinguish one SBR as having an appreciable fraction of long sequences of styrene units.

Nuclear magnetic resonance data can provide detailed sequencing information that can help relate polymer structure to the particular polymerization method used to prepare solution SBRs. This can be demonstrated for the series of experimental medium vinyl solution SBRs shown in Table 6. Using BuLi complexed with TMEDA, in a batch polymerization of butadiene with styrene in cyclohexane at 65°C, the amount of 1,2-structure was maintained at about 45% and the styrene content in the polymer was varied from 20 to 50 wt%. For the medium-vinyl SBR (MVSBR) containing 20% or less styrene, the units of styrene are incorporated predominantly as isolated, diads and a small fraction of triads (in an amount that decreases with decreasing styrene content). As styrene content increases from 25 to 50 wt%, an increasing fraction of long sequences of styrene units is produced by this system. The last entry in Table 6 demonstrates that doubling the TMEDA concentration at constant BuLi raises the vinyl content as expected, and totally eliminates the long sequences of styrene in the polymer containing 50 wt% styrene. Thus, both the amount of styrene and TMEDA are crucial polymerization parameters that determine the particular styrene sequence distribution of solution SBRs.

2.6. Macrostructure Variations
In addition to polymer composition and microstructure, molecular weight, molecular weight distribution (MWD) and chain architecture (collectively referred to as polymer macrostructure) are considered to

be fundamental parameters affecting dynamic mechanical properties and the ability of rubber, in part, to be mixed, processed and fabricated into finished products. Anionic catalyst systems for preparing SBR and BR rubbers offer a variety of ways for effectively controlling polymer macrostructure.

As previously mentioned, anionic polymerization based on alkyllithiums involves essentially only two processes, initiation and propagation, because chain termination is absent. If all the growing chains are formed during a short time interval (i.e. initiation is fast relative to the propagation process) and if all the chains have equal opportunity to grow, then the resulting polymer will have a narrow molecular weight distribution. This feature can lead to low dynamic energy losses, which in tyre applications means low rolling resistance but greatly decreases the processability and results in viscoelastic behaviour that is characteristic of cold flow during long-term storage. The remedy for these deficiencies is usually found by altering the MWD and/or introducing a certain amount of controlled branching in the polymer molecule.

2.6.1. Broadening MWD and Branching

There are several approaches to broaden the MWD and introduce chain branches. The most commonly used are procedures that affect the initiation process,[29] and the use of branching comonomers such as divinylbenzene (DVB)[30] and chain end linking (coupling). The latter process has been widely used for coupling non-terminated chain ends with a variety of multifunctional organic and inorganic coupling agents. Examples of coupling agents are diesters, diketones, multifunctional vinyl compounds and metal halides. With respect to coupling of polymer chains bearing one reactive carbanion per chain, the number of branches will be controlled by the functionality of the coupling agent. For coupling agents that join chains by a polymer–halogen exchange reaction, the number of branches will be equal to or less than the number of halogen atoms of the coupling agent. When the coupling agent is a branching monomer such as DVB, which is capable of participating in both polymerization with itself and chain end linking, star polymers having as many as 30 branches (depending on reaction stoichiometry and conditions) have been produced.

Another approach to introduce branching is by the use of polyfunctional initiators. These initiators are based upon the reaction of an alkyllithium (*sec*-BuLi complexed with triethylamine) with a multi-functional vinyl compound (DVB). The average functionality of these multilithiated initiators, and consequently the resulting living polymer, is largely controlled by the mole ratio of *sec*-BuLi/DVB, by the rate of addition of DVB to *sec*-BuLi and by reaction temperature. A distinguishing feature of star polymers prepared with polyfunctional Li initiators is that the terminal chain carbanions can be reacted with reagents such as ethylene oxide and carbon dioxide to give functionally terminated materials. GenCorp, Lithium Corporation of America and Phillips Petroleum Company patents have reported the preparation of hydrocarbon soluble polyfunctional initiators and their use for the synthesis of such ω-reactive star polymers.[33-34] Star-shaped solution SBRs of this kind have been studied for application in passenger tyre tread compounds for improved tread properties,[35] and in applications as impact modifiers and low-profile, low-shrink additives for fibreglass-reinforced plastics.[36]

The effect of coupling a monofunctional polymer chain with a tetrafunctional metal halide on the MWD of a solution SBR is shown schematically in Fig. 6. For the case of non-quantitative coupling, as shown here, the components of the resulting bimodal MWD consist of

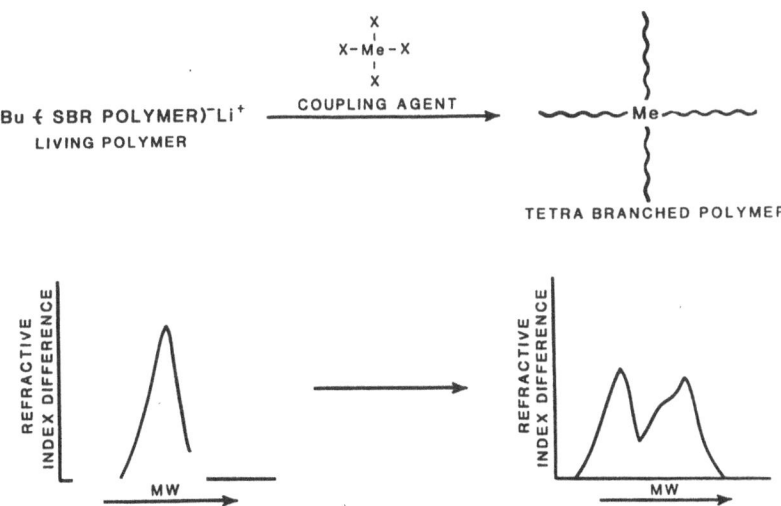

FIG. 6. Method for broadening MWD and introducing chain branching for property and processing improvements.

the linear precursor and primarily the tetrabranched polymer (e.g. Sn-coupled SBR). The formation of long chain branches can provide higher resistance to cold flow and more flow at moderate and high shear rates encountered in extrusion operations than the corresponding linear polymer of the same molecular weight.

2.6.2. Tin-Coupled Polymers

Of particular significance is the unique behaviour of Sn-coupled solution polymers. The technology of producing these Sn-coupled solution SBRs was discovered and developed for commercialization by Phillips Petroleum Company.[37] The resulting polymers are organotin compounds having a fairly weak carbon–Sn bond.[38] This bond can be ruptured easily with carboxylic acids leading to a reduced molecular weight, especially when the group attached to Sn is an allylic group as in the case of 1,4-butadienyl.[39] The extent of molecular weight reduction of Sn-coupled polymers treated with stearic acid, a commonly used component in rubber vulcanization systems, can be controlled much more effectively than is possible by breakdown of conventional elastomers by mechanical energy or by means of chemical peptizers.

Since the oil crisis in 1973, Japanese workers have conducted extensive evaluations on Sn-coupled solution SBRs in passenger tyre tread compounds for meeting the Japanese government's mandate for improving the fuel economy of tyres.[40,41] Their tyre test data derived from Sn-coupled SBRs have indicated improved tyre tread properties, especially reduced rolling resistance and improved wet traction, relative to emulsion SBRs. Recently, Japanese workers have suggested that the benefits of Sn-coupled polymers are related to their interaction with carbon black.[42,43] When cleavage of the carbon–Sn bond takes place in the presence of carbon black during mixing, there is considered to be a chemical interaction between polymer chains bearing Sn atoms, not fully bonded, and the surface of carbon black. This increased interaction is believed to result in improved carbon black dispersion and to produce increased amounts of rubber insolubilized by carbon black (bound rubber). With respect to the benefit of Sn-coupled solution SBRs for providing reduced energy loss, further studies are needed to determine the relative contributions of black interactions or, more simply, the enhanced chain linearity and molecular size uniformity that can occur as a result of carbon–Sn cleavage.

2.7. Functional Terminated Polymers

Placing functional groups on the terminal chain end(s) of anionic polymers (e.g. by conversion of active centres to —OH and —COOH groups) by reaction with ethylene oxide or carbon dioxide, followed by hydrolysis, is of interest because of the potential for further chain extension and reactions through the reactive end groups. Diene polymers with functional groups have been prepared with mono- and multi-lithium initiators, as well as with functionalized organolithium initiators.

Anionic polymerization techniques have long been employed to prepare telechelic polymers, i.e. liquid polymers with two or more terminal functional groups per polymer chain. The quality of functionally terminated polymers is strongly dependent on the initiator used; this is especially so for dilithio initiators for use in preparing telechelic and triblock ABA block polymers. The following guidelines are useful for selection of a suitable difunctional initiator for obtaining diene polymers with a high 1,4-(*cis* + *trans*)-content: sufficient solubility in a hydrocarbon solvent, purely difunctionality, resistance to thermal degradation at use temperatures, rapid initiation of polymerization, a relatively low molecular weight and low cost of preparation.

A dilithium initiator that satisfies all of the above criteria is not available commercially. However, an initiator that has many of the characteristics, with the exception of general availability, is the adduct of 1,3-bis(phenylethenyl)benzene and two equivalents of *sec*-BuLi:[44]

This initiator is moderately soluble in hydrocarbon solvent, thermally stable and can produce hydroxyl-terminated polybutadienes with functionalities of two. Additionally, studies on its use for polymerization of butadiene have shown predictable molecular weights and near monodisperse MWD.[45] However, a bimodal MWD has also been reported.[46] Quirk and co-workers have also recently investigated the use of 1,3-bis(phenylethenyl)benzene as a 'living' coupling agent (i.e. a coupling agent which retains active carbanionic centres in the coupled product) for the preparation of star polymers with compositionally heterogeneous arms.[47]

Another way to introduce an end group on a polymer chain is to use a functionally substituted organolithium initiator.[48] Schulz and co-workers have prepared lithium initiators containing acetal groups,

$$\text{Li}-\text{R}-\text{O}-\underset{\underset{\text{OC}_2\text{H}_5}{|}}{\text{CH}}-\text{CH}_3$$

which are stable to attack by RLi. After the polymer is prepared, the protective acetal moiety is hydrolysed to provide a terminal hydroxyl group. Dihydroxyl polybutadiene polymers can be prepared with this approach by coupling the acetal polybutadienyllithium with dichloro-dimethyl silane.[49] Another option is to terminate the dienyl lithium active centre with a compositionally different reactive functional group, e.g. CO_2, to produce heterogeneous functionality on the same polymer chain.

In general, chain end functionalization is applied to relatively low molecular weight polymeric intermediates, thus having a sufficiently high concentration of reactive end groups for further reactions, e.g. chain extension or crosslinking reactions useful in network formation from telechelic polymers. Functional groups have recently been attached to the chain ends of high molecular weight ($\bar{M}_n > 100\,000$) solution polymers for use in tyre tread compounds.[50] Large increases in the resilience at 60°C (a measurement related to rolling resistance) of tread vulcanizates containing high-vinyl polybutadienes have been attributed to the adduct of polybutadienyl Li active centres with 4,4'-bis(diethylamino)benzophenone:[51]

$$(\text{CH}_3\text{CH}_2)_2-\text{N}-\underset{}{\bigcirc}-\overset{\overset{\text{O}}{\|}}{\text{C}}-\underset{}{\bigcirc}-\text{N}-(\text{CH}_2\text{CH}_3)_2$$

Increased dispersibility of carbon black by the chemically modified polymer is considered to relate to the improved dynamic properties, although the exact mechanism is not clear.

This area of research is expected to grow in importance as polymerization chemists search for specific reactive end groups that can lead to further improvements in performance features of rubbers for application in tyre treads.

3. HIGH-*TRANS* SBR (HTSBR)

3.1. Introduction to HTSBR

The synthesis, structure and properties of a new class of crystallizing butadiene (Bd) elastomers will be described in this section. These butadiene-based rubbers crystallize because they have a sufficiently high percentage of the *trans*-1,4-isomer in the polymer chain (greater than 75%). The copolymers of butadiene and styrene having a preponderance of *trans*-1,4-butadiene linkages are referred to as High-Trans Styrene–Butadiene Rubbers, or 'HTSBR'. For the first time, crystallizable solution SBRs can be prepared with all the versatility that is characteristic of alkyllithium polymerization. They are unlike the HTBRs prepared with Ziegler–Natta catalysts in that copolymers with styrene are readily obtained and differ from Alfin SBRs (75% *trans*, 25% vinyl) in having only 2–4% vinyl content, one-tenth of that of the Alfin polymers, resulting in a much lower glass transition temperature. A further distinguishing characteristic of these HTSBRs is the linearity of the polymer chains.

A trimetallic catalyst of Ba/Mg/Al used in the synthesis of HTSBR will be described and its 'living polymerization' characteristics will be considered in some detail. Copolymerization reactivity ratios and styrene sequence distribution of HTSBRs will be compared with emulsion SBRs. This section will conclude with the crystallization behaviour of HTSBRs. Technological properties will be discussed in Sections 4.6 and 4.7.

3.2. HTSBR Synthesis

The ternary catalyst system, shown in Fig. 7, consists of Ba *tert*-butoxide [Ba(t-BuO)$_2$] in combination with a complex of dibutyl-magnesium (Bu$_2$Mg) and triethylaluminium (Et$_3$Al): Ba/Mg/Al. It was found to be useful in controlling stereoregularity of butadiene polymerizations and in providing SBRs with high tack and green

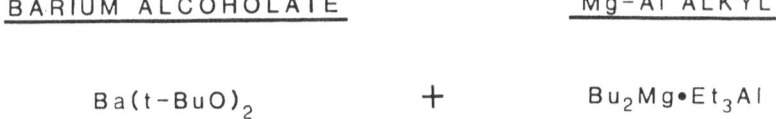

BARIUM ALCOHOLATE   Mg–Al ALKYL

Ba(t–BuO)$_2$ +   Bu$_2$Mg•Et$_3$Al

FIG. 7. Components of Ba/Mg/Al polymerization catalyst.

strength. Although dibutylmagnesium and its complex with triethylaluminium are not suitable as the sole initiator to polymerize butadiene and styrene, their complex with barium *tert*-butoxide in a hydrocarbon medium can initiate polymerization of both. Polymerizations are typically carried out in aliphatic or cycloaliphatic hydrocarbon solvents, with approximately 15–20 wt% monomer concentrations.

A fairly narrow MWD results from this polymerization. The living nature of polymerization allows control of branching and the production of rubbers with broadened MWDs, as shown in Fig. 8, for the case of post-reaction of reactive polymer carbanion chain ends with DVB. These branched polymers are easily oil-extended and exhibit lower cold flow than their linear counterparts.

The most important polymerization variables which affect the molecular structure of polybutadienes, prepared by the Ba/Mg/Al catalyst, are the ratio of barium salt to Bu_2Mg (Ba^{2+}/Mg^{2+}), the polymerization temperature and catalyst concentration. The effect of these variables on *trans*-content and molecular weight is summarized in Table 7. Whilst the trends are shown for butadiene homopolymerization, essentially equivalent responses have been obtained for copolymerizations of styrene and butadiene.

The dependence of polybutadiene microstructure on Ba^{2+}/Mg^{2+} mole ratio is shown in Fig. 9, at a fixed Mg/Al ratio of 6/1 and for polymerizations in cyclohexane at 50°C. *Trans*-content is seen to

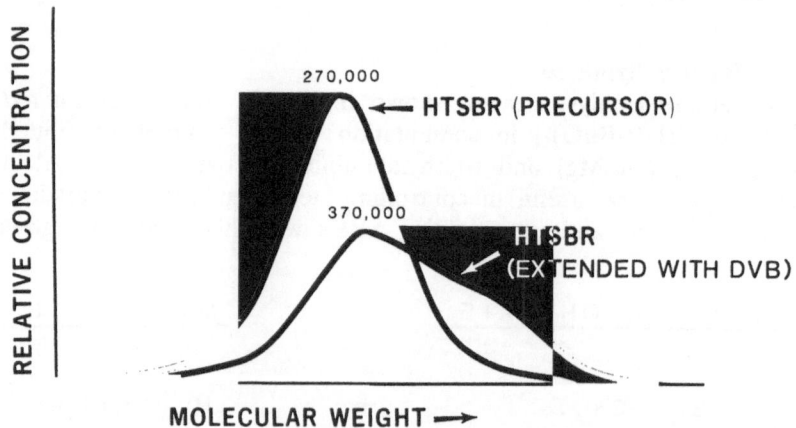

FIG. 8. Effect of chain extension on the molecular weight distribution of HTSBR.

TABLE 7
EFFECT OF MAIN POLYMERIZATION VARIABLES ON
POLYMER STRUCTURE AND MOLECULAR WEIGHT

Variable	trans-1,4-Polybutadiene portion (%)	Molecular weight
Increased polymerization temperature	Decrease	Decrease
Increased catalyst concentration	Increase	Decrease
Increased Ba[a]	Decrease	Decrease
Increased Al[a]	Increase	Increase

[a] Constant R_2Mg.

increase to a value of about 90% for low levels of Ba salt. Vinyl content is low, decreasing from about 7% to as low as 2% with decreasing Ba salt. The rate of polymerization is influenced by the Ba^{2+}/Mg^{2+} mole ratio, e.g. polymer is not formed below a Ba^{2+}/Mg^{2+} mole ratio of 0·05.

The dependence of trans-content on polymerization temperature is shown in Fig. 10. The amount of trans-1,4-structure is increased from 79 to 90% as polymerization temperature is reduced from 80 to 30°C. Concurrently, the vinyl and cis-1,4-contents decreased.

Increasing catalyst concentration, expressed as Bd/MgR_2, gives polymers with higher trans-contents, although lower molecular weights, as shown in Fig. 11. It is considered likely that the combination of high catalyst concentration and low polymerization temperature is conducive to the formation of a specific Ba/Mg/Al complex.

The molecular weight of HTSBR is not only a function of the ratio of monomer to Bu_2Mg, but it also depends on the ratio of Ba/Mg, as shown in Fig. 12. In this figure, $\bar{M}_s$ is the stoichiometric molecular weight calculated from the ratio of grams polymer formed/moles of active Bu_2Mg. For equimolar amounts of Ba and Mg, the number-average molecular weight, $\bar{M}_n$, is about one-half of $\bar{M}_s$. At lower ratios of Ba/Mg, $\bar{M}_n$ increases at constant $\bar{M}_s$. The dependence of molecular weight on Ba/Mg mole ratio suggests that the degree of participation of alkyl–Mg bonds in initiating polymer chains increases with Ba

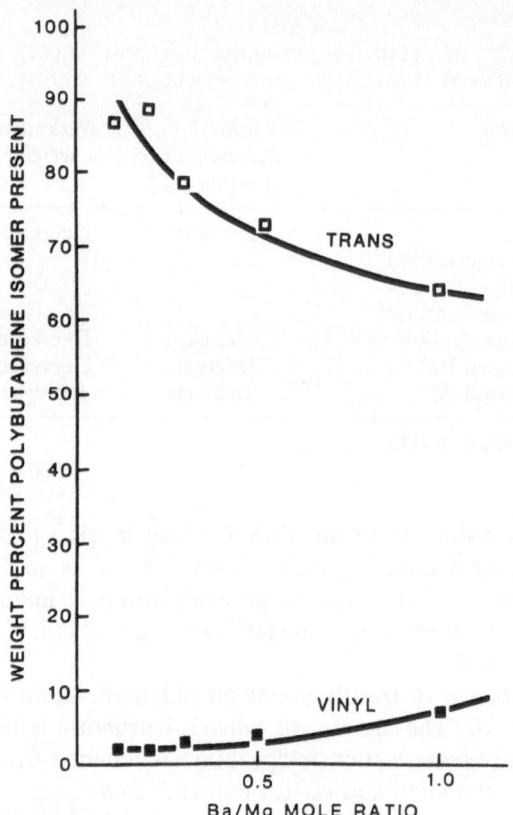

FIG. 9. Polybutadiene microstructure versus the mole ratio of Ba/Mg in the Ba/Mg/Al catalyst. (Reproduced with permission from ref. 24, by courtesy of the American Chemical Society, Washington, DC.).

content. Apparently, Ba can cause both carbon–Mg bonds to partici-pate in butadiene polymerization at Ba/Mg = 1.

3.3. Living Polymers

The absence of any fortuitous chain termination step is an established feature of homogeneous anionic polymerization of diene and styrene in hydrocarbon solvent with organolithiums. Similarly, in the absence of adventitious chain terminating agents, the chain ends formed with Ba/Mg/Al catalysts are stable to an appreciable extent.

The lifetime of these chain ends of polybutadiene and polystyrene is

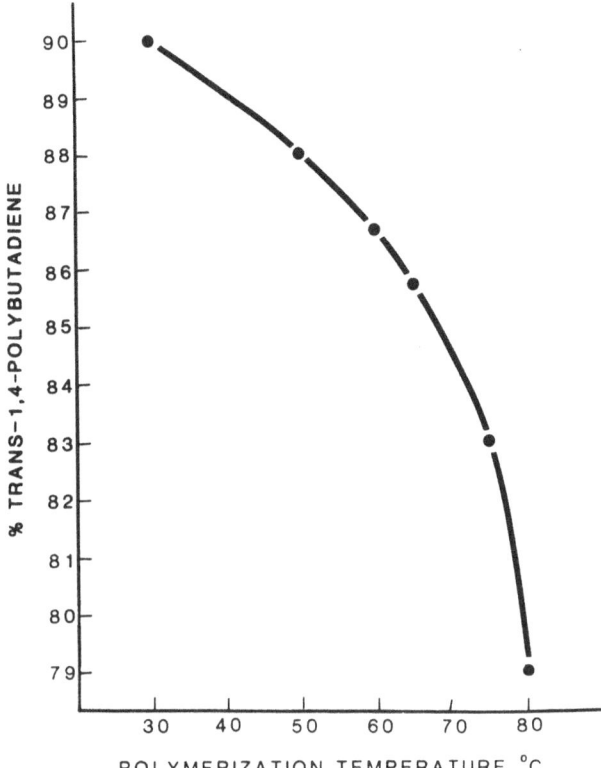

FIG. 10. Variation of percentage of *trans*-1,4-content with polymerization temperature. (Reproduced with permission from ref. 24, by courtesy of the American Chemical Society, Washington, DC.)

sufficiently long to exhibit an increase in molecular weight with both an increase in the extent of conversion as well as with the addition of more monomer. The living nature is further demonstrated by prepara-tion of AB-type polystyrene–polybutadiene diblock copolymer using a two-stage sequential polymerization process. A comparison of MWD curves (Fig. 13), their relative peak positions and shapes, demon-strates the successful preparation of a diblock copolymer. End-linking the non-terminated chains of polybutadiene of the above polystyrene-b-polybutadiene with DVB results in star-shaped block copolymers that behave as thermoplastic elastomers.

The reactive anionic end groups of polybutadiene prepared with the

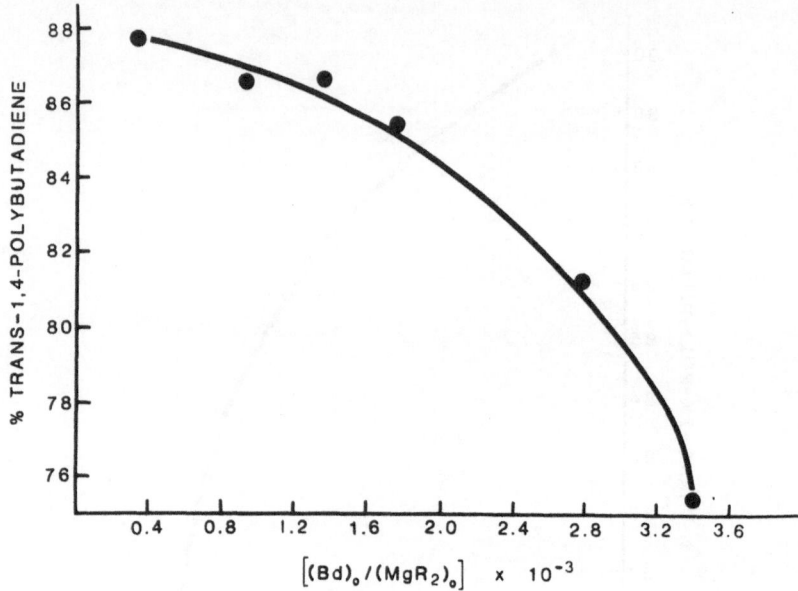

FIG. 11. Variation of *trans*-1,4-content of polybutadiene with molar ratio of butadiene to dialkylmagnesium. (Reproduced with permission from ref. 24, by courtesy of the American Chemical Society, Washington, DC.).

Ba/Mg/Al catalyst system discussed above can be readily converted into a variety of functional end groups. One of the most useful and widely used functionalization reactions is hydroxyethylation using ethylene oxide, as shown in Fig. 14. Here, a mono-hydroxyl-terminated high-*trans* polybutadiene was prepared by adding a small amount of ethylene oxide to a solution of carbanions, followed by hydrolysis. This particular example demonstrates the ability to attach hydroxyl groups molecularly to a crystalline polybutadiene. The use of a bifunctional diorganomagnesium could provide a very useful tele-chelic polymer for further extension to form a rubber network.

The nature of the chain ends of HTBR has been further investigated by using a radiochemical method[52] to determine the number of active polymer carbanions formed with the Ba/Mg/Al catalyst. The basis of this method is the quantitative reaction of metal–polymer bonds with tritium atoms of tritiated alcohol, as shown below:

Polymer—metal + RO—^{3}H → polymer—^{3}H + RO—metal.

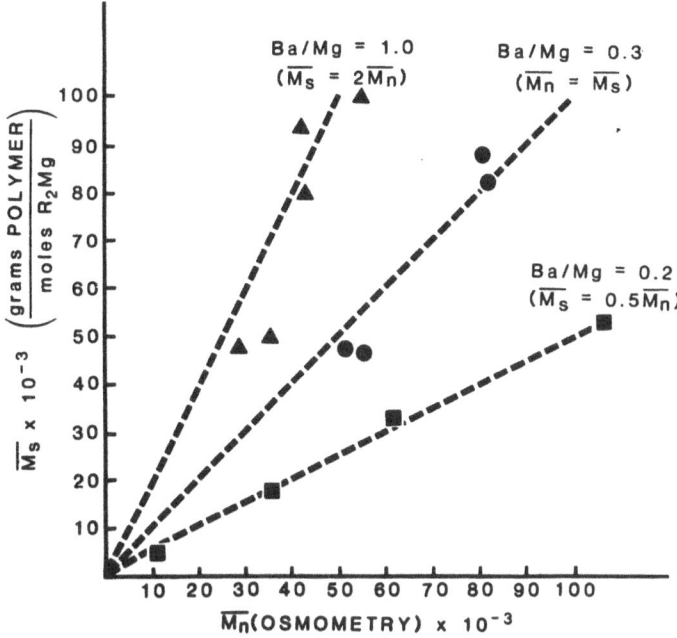

FIG. 12. Effect of mole ratio of Ba/Mg on relationship of polymer molecular weights.

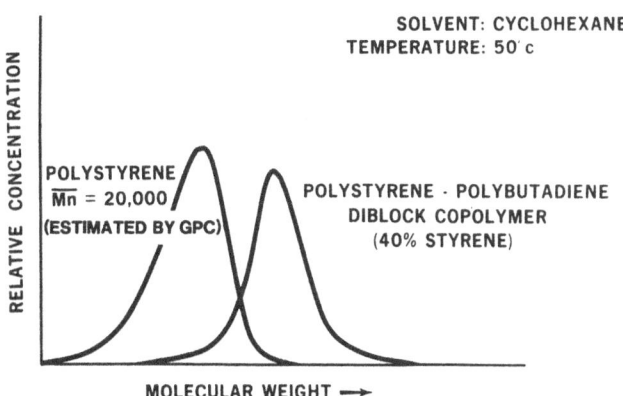

FIG. 13. Gel permeation chromatograms of polystyrene and polystyrene–polybutadiene diblock copolymer prepared with Ba/Mg/Al catalyst. (Reproduced with permission from ref. 24, by courtesy of the American Chemical Society, Washington, DC.).

$$\text{\textasciitilde\textasciitilde\textasciitilde} CH_2-CH=CH-CH_2^-Me^+ \quad + \quad CH_2-CH_2 \quad \xrightarrow{H_2O}$$
$$\underset{O}{\diagdown \diagup}$$

$$\text{\textasciitilde\textasciitilde\textasciitilde} CH_2-CH=CH-CH_2-CH_2-CH_2-OH \quad + \quad MeOH$$

FIG. 14. Chain end derivatization. Hydroxyl functionality $= 0.91$.

Using this method, we have compared equivalent molecular weights (based on tritium content) of HTBR with their $\bar{M}_n$ values measured by membrane osmometry, as shown in Fig. 15. The dashed line in the Figure, with a slope of unity, represents equivalency of both molecular weight measurements, for the case where all of the organometallic sites retain their capacity to react with tritiated alcohol. The data demonstrate that the Ba/Mg/Al system lacks spontaneous termination reaction. This is true to a very large extent, although there does appear to be some falling-off of active chain end content in the high molecular weight region.

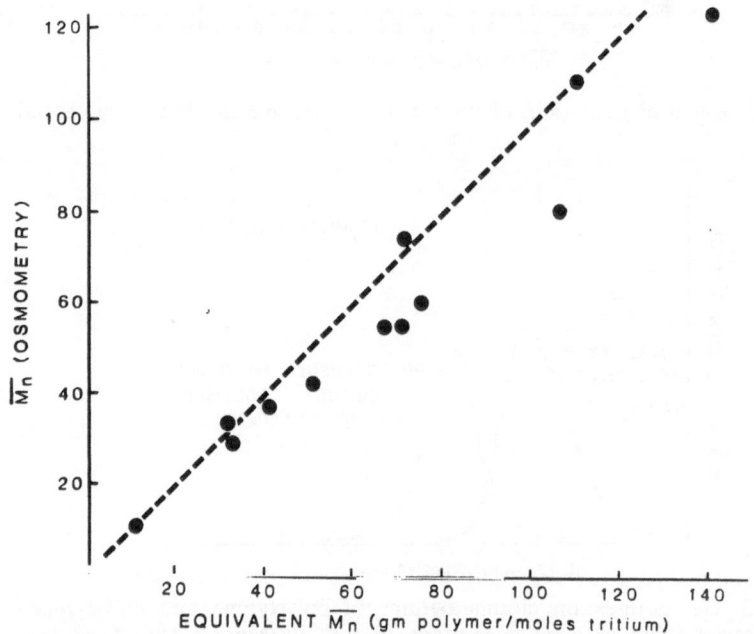

FIG. 15. Comparison of equivalent weight and number-average molecular weight of *trans*-butadiene polymers.

3.4. Copolymerization of Butadiene and Styrene

Random copolymers are commercially produced by both emulsion and solution polymerizations. It was discovered in the author's laboratory that the use of barium alkoxides in combination with organometallic compounds is yet another method to prepare random styrene/butadiene solution rubbers. With Ba/Mg/Al catalyst systems, random SBRs are obtained with no more than 4% vinyl unsaturation. The principal effect of lower vinyl content is a lower glass transition temperature than that of a corresponding, conventional solution SBR.

It is useful to compare the detailed monomer sequence distribution in SBRs obtained with emulsion free-radical and *n*-BuLi systems with the SBRs prepared with Ba/Mg/Al. We will begin by discussing how styrene incorporation varies with extent of conversion, then examine reactivity ratios and sequence distribution information obtained by proton NMR.

3.4.1. Variation of Composition with Conversion

The amount of styrene incorporated into an experimental high-*trans* SBR at any given conversion is compared (Fig. 16) with that for a polymer prepared by free-radical emulsion polymerization and one

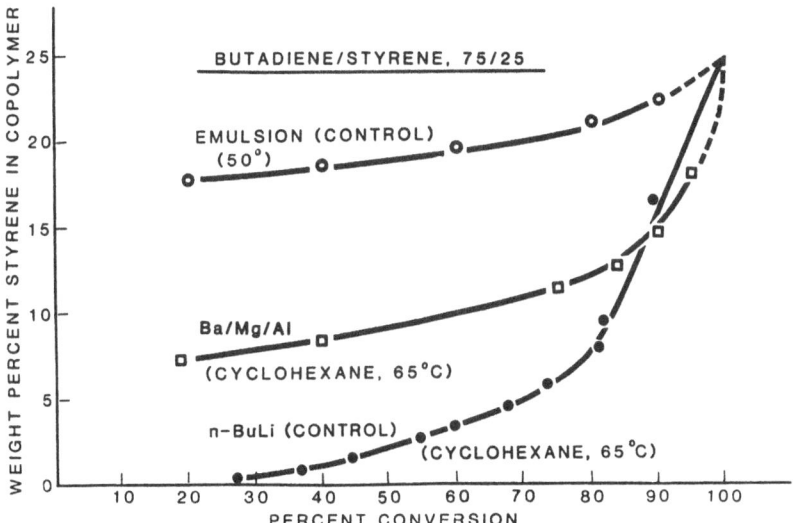

Fig. 16. Copolymer composition variation with percentage conversion.

prepared using solution polymerization initiated with unmodified n-BuLi in cyclohexane. The initial charge ratio is styrene/butadiene, 25/75. Styrene content in polymer produced with the Ba/Mg/Al catalyst system, at conversions of 20 to 85%, is intermediate between that for the nearly totally random emulsion SBR and the AB-type butadiene–styrene graded block copolymer obtained with n-BuLi. In the high-conversion region (85–100%), the monomer pool is increasingly richer in styrene than in butadiene and the remaining increments of polymer formed with both n-BuLi and Ba/Mg/Al become increasingly high in styrene content. However, the analysis of HTSBR samples, polymerized to high conversion, by oxidative degradation with OsO$_4$ and tert-butyl hydroperoxide[25] produced no measurable amounts of polystyrene.

3.4.2. Reactivity Ratios of Styrene–Butadiene Copolymer Reactions

Greater insight into the copolymerization behaviour of butadiene and styrene with a Ba/Mg/Al catalyst system can be seen by examining their reactivity ratios. We used the graphical method of Kelen and Tüdos[53] to determine the r-values, given in Table 8. For comparison, r-values are given for emulsion SBR prepared at 60°C and for a graded solution SBR copolymer prepared in cyclohexane with butyllithium. For HTSBR, the particular set of obtained reactivity ratios, $r_{Bd} = 4 \cdot 4$ and $r_{Sty} = 0 \cdot 5$, indicates that a butadienyl carbanion is about four times more likely to add butadiene than to add styrene, and a styryl carbanion is two times more likely to add butadiene than to add itself. This means that as long as the monomer feed concentration is rich in

TABLE 8
COMPARISON OF REACTIVITY RATIOS FOR STYRENE/BUTADIENE COPOLYMERIZATIONS

Catalyst	r_{Bd}	r_{Sty}
Free radical[a]	1·39	0·78
Butyllithium[b]	15·5	0·04
Ba/Mg/Al[c]	4·40	0·52

[a] Emulsion polymerization, 60°C.
[b] In cyclohexane, 25°C; Morton, M., Anionic polymerization: Principles and Practice, Academic Press, New York, 1983.
[c] In cyclohexane, 60°C.

butadiene, there is a tendency for long sequences of butadiene, regardless of the chain end type. It is apparent that this tendency is more pronounced for butyllithium and reduced in the free-radical emulsion system. Additionally, an increase in Ba/Mg mole ratio will alter the r-values, indicating a decreased tendency for producing long runs of butadiene and an increased degree of randomness; however, this takes place at the expense of reduced $trans$-1,4-content.

3.4.3. Characterization of Styrene Sequence Distribution by Proton NMR

As previously mentioned, proton NMR has proved to be a valuable tool for obtaining sequence distribution data. This is demonstrated here by comparing the relative differences in styrene sequencing in HTSBR and an emulsion SBR, at equivalent styrene content (15 wt%). The shaded area between the proton NMR (80 MHz) curves from HTSBR and E-SBR, shown in Fig. 17, represents the additional amount of longer styrene sequence lengths present in HTSBR (top curve in right-hand portion of curve) versus emulsion SBR (bottom curve). Results from high-resolution NMR (200 MHz) indicate that a major portion (80% +) of styrene placements in HTSBR are random (styrene present collectively as isolated units, diads and triads). However, HTSBR does contain a small fraction of styrene sequences varying in length from four to seven units, relative to less than 1% of such units in E-SBR. Long sequences of styrene of length greater than eight have not been observed in HTSBR.

3.5. Crystallinity in HTSBRs

3.5.1. Properties of Uncrosslinked HTSBR

The ratio of styrene to butadiene in HTSBR produces significant effects on the properties of the uncrosslinked polymer. $trans$-1,4-Polybutadiene (90% $trans$-content), which is produced by the Ba/Mg/Al catalyst system, is a thermoplastic material having a glass transition temperature of about $-90°C$, as measured by DSC, as shown in Fig. 18. The polymer shows two pronounced crystalline melting temperatures at 50°C and 88°C, which are characteristic of the two pseudohexagonal crystalline forms (Form I and Form II) of $trans$-1,4-polybutadiene.[54] Since, as we have seen, styrene is distributed in a predominantly random fashion in HTSBR, the average sequence lengths of $trans$-polybutadiene segments can be shortened

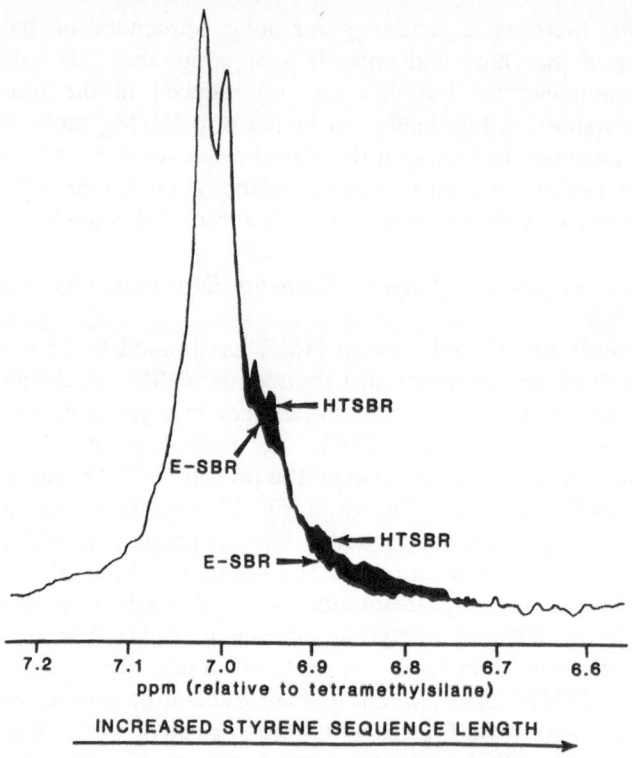

FIG. 17. Comparison of proton NMR absorption patterns of HTSBR and E-SBR.

with increasing amounts of styrene, as well as by decreasing the level of *trans*-content. The result is a depression of the crystalline melting temperature.[55] Figure 19 shows the DSC trace of an HTSBR with 85% *trans*-1,4-placements of butadiene and containing 15 wt% styrene. A broad melt endothernm at 11°C, followed by two smaller endothermic transitions at 33°C and 44°C, can be seen. Also observed is a glass transition temperature (T_g) at −83·5°C which is intermediate between the T_g values for polybutadiene and polystyrene homopolymers. For this copolymer, a T_g value of −75°C is predicted from the Fox equation[56] for random copolymers. The somewhat lower observed T_g is considered a result of the particular sequence distribution of this HTSBR, discussed in Section 3.4.

As mentioned above, the crystalline melt temperature of HTSBR

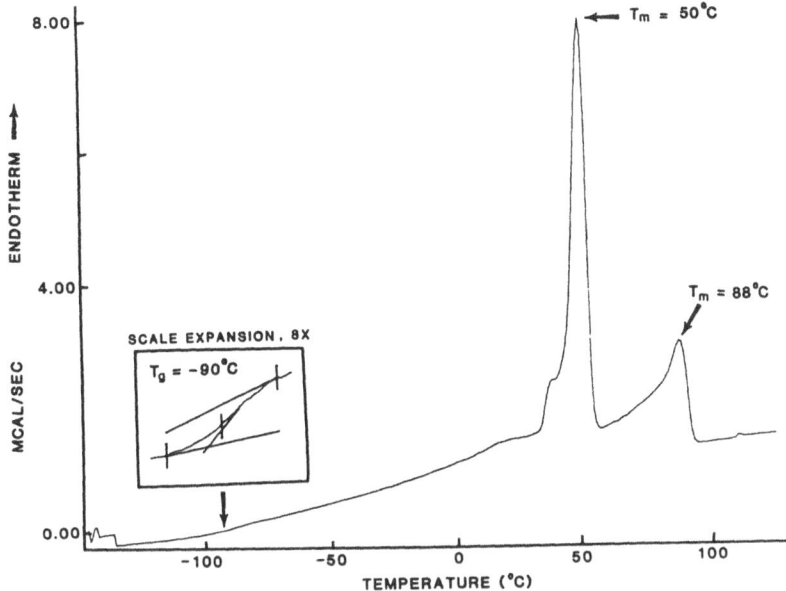

FIG. 18. DSC curve of 90% *trans*-1,4-polybutadiene prepared with Ba/Mg/Al. (Reproduced with permission from ref. 24, by courtesy of the American Chemical Society, Washington, DC.).

can be varied by adjustment of the *trans*-1,4-content and/or the incorporation of styrene. The regulation of T_m by changes in *trans*- and styrene contents is shown in Fig. 20. A crystalline melt temperature near room temperature is obtained in an HTSBR with about 15 wt% styrene and 85% *trans*-structure. A T_m near room temperature in an SBR offers important technological property advantages over general-purpose type SBRs produced by free-radical emulsion and anionic solution polymerization, and will be discussed in Section 4.7.

So far, we have examined the crystallinity in HTSBR polymers in the unstrained, uncrosslinked state. There are few, if any, synthetic rubbers that are amorphous, or have only a low amount of crystallinity at room temperature, but undergo crystallization on stretching. Those that do can exhibit the levels of green strength and tack strength that are highly desired in tyre construction and are characteristic of natural rubber (NR). Information on strain-induced crystallization behaviour of HTSBR is important to the understanding of these performance properties.

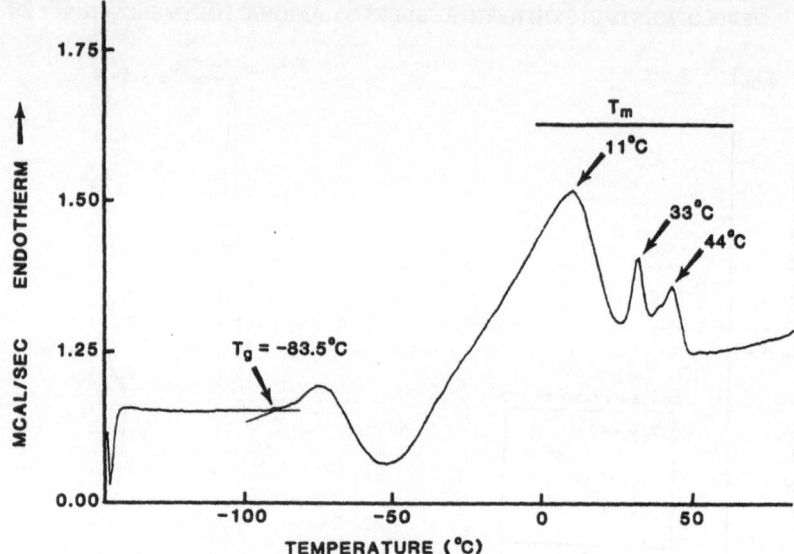

Fig. 19. DSC curve of high-*trans* SBR (15% styrene, 85% *trans*). (Reproduced with permission from ref. 24, by courtesy of the American Chemical Society, Washington, DC.).

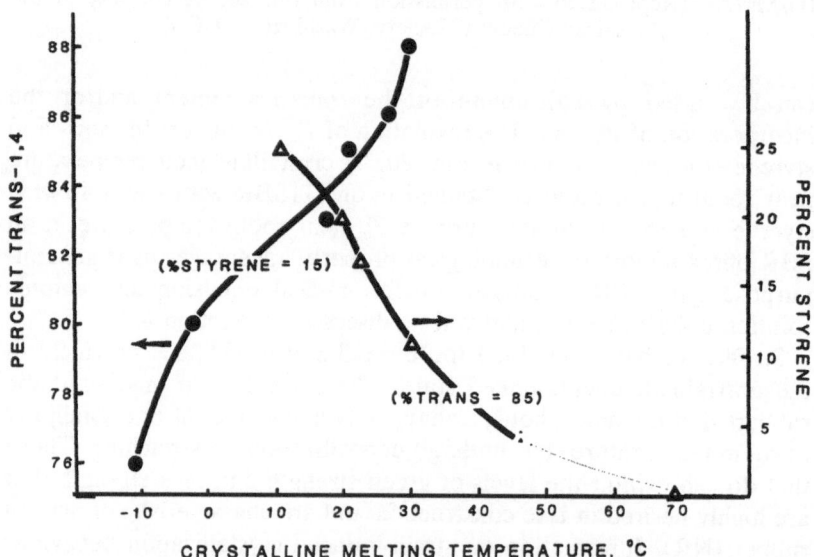

Fig. 20. Variation of melting temperature of HTSBR with *trans*-1,4- and styrene contents.

3.5.2. Strain-Induced Crystallization Behaviour

One of the useful techniques for studying the strain-induced crystal-lization behaviour of rubbers is rheo-optical measurements.[57] By measuring the variations in stress relaxation and birefringence that take place when a rubber is stretched as a function of temperature, one can characterize the crystallization behaviour with respect to elongation and temperature. Figure 21 shows the percentage crystal-linity obtained by birefringence measurements for an unfilled vulcaniz-ate of NR at various elongations as a function of temperature. The relative shapes of the curves in this Figure show the pronounced temperature and strain dependence on the strain-induced crystal-lization of NR. Of particular importance are the relatively high amounts of crystallinity that develop at room temperature. Figure 22 shows the percentage crystallinity as a function of temperature and extension ratio for HTSBR (22% styrene, 87% *trans*) prepared with a Ba/Mg/Al catalyst. A comparison of Figs 21 and 22 shows that the

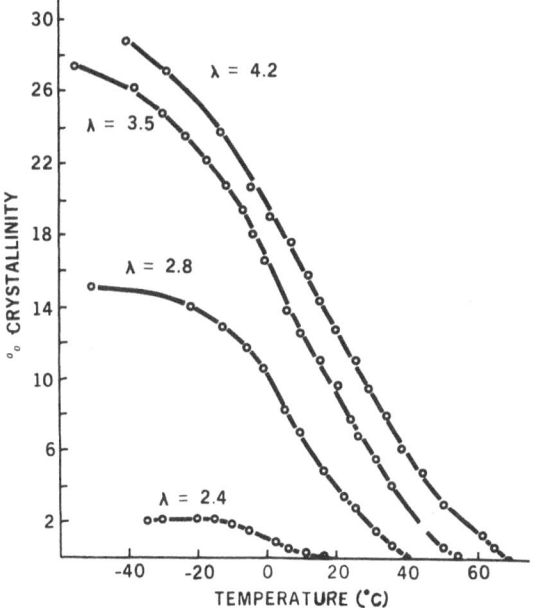

FIG. 21. Percentage crystallinity of NR as a function of temperature and extension ratio. (Reproduced with permission from ref. 24, by courtesy of the American Chemical Society, Washington, DC.).

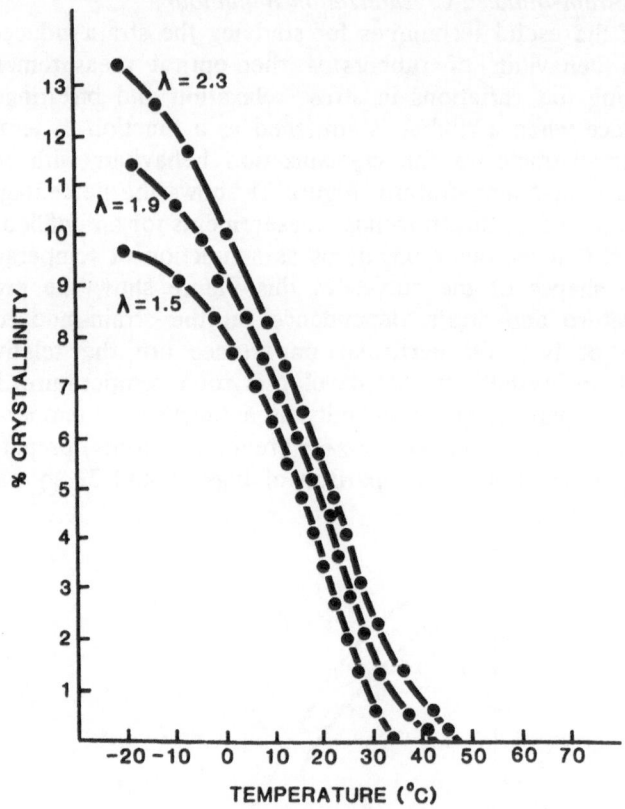

Fig. 22. Percentage crystallinity of HTSBR (22% styrene, 87% *trans*) as a function of temperature and extension ratio. (Reproduced with permission from ref. 24, by courtesy of the American Chemical Society, Washington, DC.).

amount of crystallinity developed in HTSBRs is temperature-sensitive, but not as strain-sensitive as that of NR. The Ba/Mg/Al-catalysed styrene–butadiene copolymer shows a marked improvement in both strain sensitivity and in the amount of strain-induced crystallinity over HTSBRs prepared with a Ba/Li catalyst.[24]

In observing the time-dependent changes in birefringence and stress-optical coefficient, for elongated samples at 25°C, it was found that the rate of crystallization of HTSBRs was very much faster, some ten times more rapid, than that for NR.[58] This is consistent with the reported rates of isothermal crystallization for NR (2·5 h at −26°C) and for 80% *trans*-1,4-polybutadiene (0·3 h at −3°C) in the relaxed state.[59]

The main conclusions concerning the strain-induced crystallization behaviour of high-*trans*-polybutadiene-based rubber and natural rubber are: (1) the rate of crystallization is extremely rapid compared with that of NR; (2) the amount of strain-induced crystallization is small compared with that of NR, especially at room temperature; and (3) for the HTSBRs relative to NR, crystallization is more sensitive to temperature at low extension ratios, and crystallization is less sensitive to strain.

4. TECHNOLOGICAL PROPERTIES OF SOLUTION RUBBERS

4.1. Introduction

Prior to the interest in fuel savings, the challenge for improved tyre tread compositions was one of obtaining a compromise between traction and abrasion resistance. Tyre tread formulations containing general-purpose emulsion SBRs and solution SBRs with polybutadiene provided a good combination of the two performance properties, namely wear resistance and traction. In addition, such tyre tread formulations exhibited excellent processability. Since the energy crisis of the early 1970s, and as a consequence of the Corporate Average Fuel Economy (CAFE) requirements, the demand for fuel savings has led tyre and rubber producers to seek improvements in rolling resistance of tyre tread compounds, since the tyre tread accounts for 30–50% of the energy dissipated during rolling deformation of a typical passenger car tyre. Thus, an important goal for rubber technologists during recent years has been to reduce rolling resistance without sacrifice in other properties. In meeting requirements for reduced rolling resistance of tyres, solution rubbers have attracted considerable interest because of the unique polymer structure control that the solution anionic polymerization process provides, as compared with, for example, emulsion and Ziegler–Natta type polymerization processes.

4.2. Structural Features Influencing Rolling Resistance, Wet Traction and Wear

The broad range of control of solution polymer structure and macrostructure of styrene–butadiene rubbers that is only possible using lithium catalysts was discussed in Section 2. We have seen how the microstructure of the butadiene units in the chain and comonomer sequence distribution can be controlled with the addition of polar modifiers and/or variations in polymerization process variables. Additionally, the unique control of macrostructure features and the new possibilities offered by reactive functional groups were discussed as part of the molecular engineering capabilities of solution anionic polymerizations.

4.2.1. Glass Transition Temperature (T_g)

The general types of solution rubbers which are currently available either commercially or in trial quantities can be described broadly in terms of their styrene and vinyl contents and, consequently, T_g, as given in Table 9. Also shown are the various types of MWDs and

TABLE 9

GENERAL TYPES OF SOLUTION POLYMERS FOR TYRES

(1) *Molecular Structure*

Polymer	Range of values		
	Styrene(%)	Vinyl(%)	T_g(°C)
Low-vinyl SBRs (LVSBR) and LVBR	0–30	8–15	−98 to −55
Medium-vinyl SBRs (MVSBR) and MVBR	0–25	15–50	−92 to −40
High-vinyl SBRs (HVSBR) and HVBR	0–15	50–75	−60 to −20

(2) *MWDs and Chain Architectures*

Shapes of MWD	$\bar{M}_w/\bar{M}_n$ (by GPC)	Chain branching agent	Chain architecture
Unimodal	>2·5	DVB	Random branching
Bimodal	2·0–2·5	Diesters	Tetra branched
Bimodal	1·5–2·0	SnCl₄	Tetra branched (labile bonds)

chain architectures characteristic of many solution polymers. A distinguishing characteristic of such solution polymers is their higher number-average molecular weights, lower heterogeneity indices and distinctly different chain architectures relative to emulsion SBR, e.g. E-SBR 1551.

In order to describe how the various molecular structural features of solution rubbers influence rolling resistance and traction, it is necessary to relate these performance properties of tyres in terms of the fundamental material properties of rubbers (Fig. 23). The loss tangent of a tread rubber vulcanizate determined at about 100 Hz and at a temperature of 60°C is a significant material property that relates to the energy dissipated per cycle in a rolling tyre, i.e. the tyre's rolling resistance.[60] The T_g of the rubber is a suitable physical material

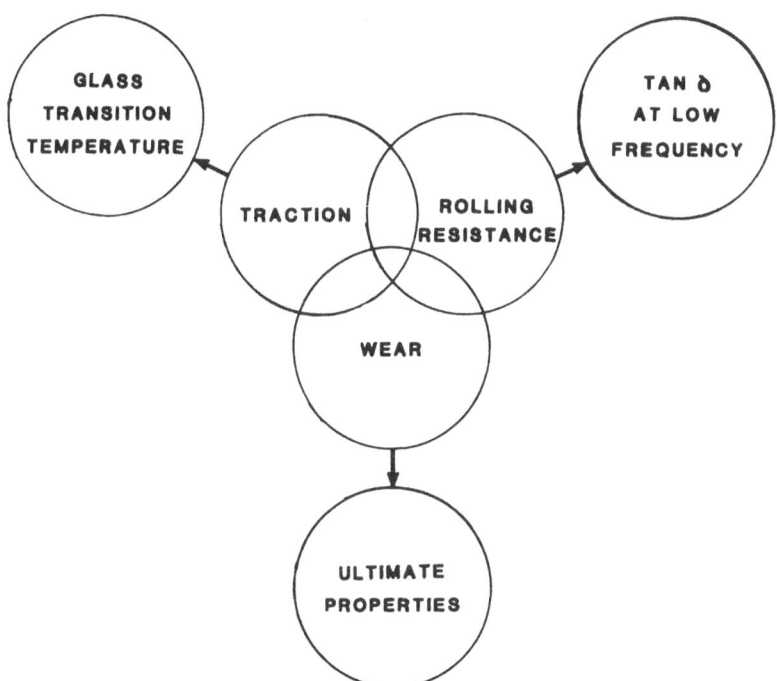

FIG. 23. Relationship of performance and fundamental material properties of tyre tread rubbers. (Reproduced with permission from ref. 62, by courtesy of Plenum Publishing Corporation, New York.).

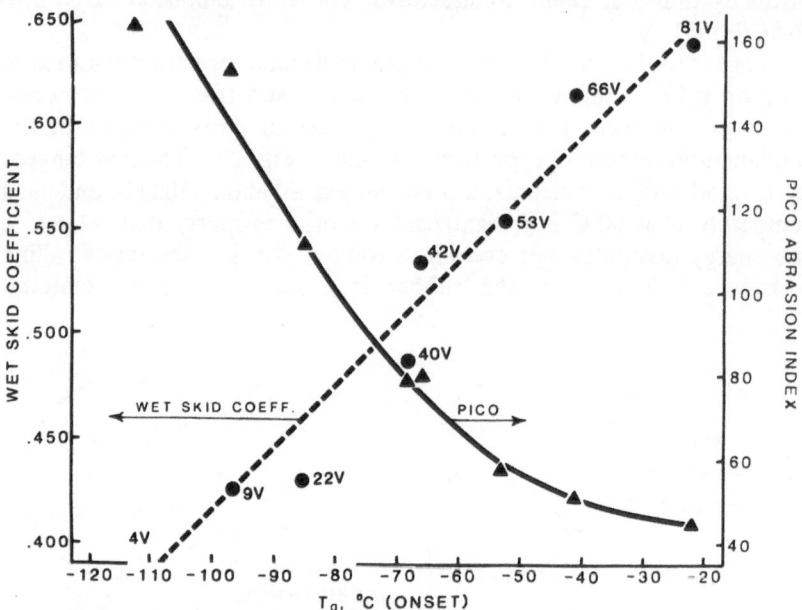

Fig. 24. Wet skid coefficient and Pico Abrasion Index versus T_g of polybutadienes.

property related to traction on a wet road during braking. Additionally, T_g can be correlated to the abrasion resistance (Pico Index) as well as wet skid behaviour, as shown in Fig. 24. In this Figure, vinyl contents (V) are shown next to data points (●). These results are for tread vulcanizates containing polybutadienes prepared with butyllithium and various amounts of TMEDA, and hence variable vinyl contents. Of course, it is possible to adjust the T_g of solution SBRs by variations in both styrene content and polybutadiene microstructure. For SBRs, the T_g of a random copolymer can be calculated from the T_g values of the corresponding homopolymers, according to the Fox equation.[56] Thus, a desired value of T_g can be obtained from many combinations of styrene, vinyl, *cis*- and *trans*-structures and, accordingly, the resulting T_g will govern the abrasion–wet traction balance.

4.2.2. Relationship between Rolling Resistance and Wet Traction
Although T_g is very useful in describing the viscous nature of vulcanized rubber, it is limited in describing the dissipation of energy

at lower deformation frequencies that may be related to rolling resistance. These relationships can be seen in a typical viscoelastic curve (Fig. 25) of a tyre tread composition in terms of the frequency dependence (or its equivalency in temperature dependence) of tan δ.

The general shape of this curve is characteristic for an elastomeric network. The relative magnitude of tan δ at any given frequency is a function of the rubber structure and its macrostructure, as well as the type, level and distribution of carbon black and other ingredients, and their specific interactions with rubber. The molecular motions of short chain segments, which govern T_g, are associated with the high-frequency (low-temperature) region of the viscoelastic spectrum. The particular segmental motions that give rise to high energy losses (high tan δ) are those that contribute to high wet grip performance. Molecular motions involving longer segments of the polymer molecule can be associated with the low-frequency (high-temperature) end of

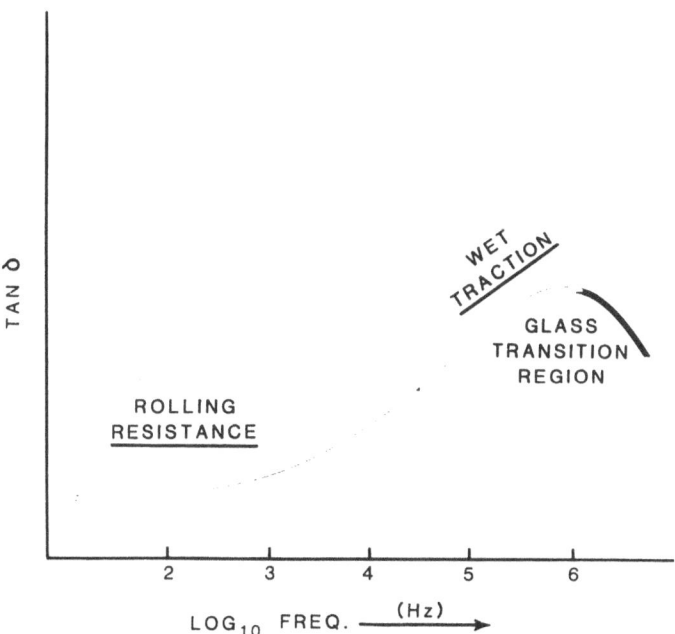

FIG. 25. Relationship of rolling resistance and wet traction to the viscoelastic loss spectrum. (Reproduced with permission from ref. 62, by courtesy of Plenum Publishing Corporation, New York.).

the spectrum. This is where low values of tan δ are needed to reduce rolling resistance. The strategy is to develop a tread rubber vulcanizate that has optimum values of tan δ measured at the appropriate frequencies and temperatures.

4.2.3. Relationship of T_g and tan δ for Conventional Tread Rubber Vulcanizates

Initially, when most conventional rubbers were evaluated for satisfying the CAFE requirements for rolling resistance, the dilemma was that when wet traction was improved (increase in T_g), there was a corresponding increase in rolling resistance, and vice versa. Figure 26 shows the relationship between T_g and tan δ for the conventional rubbers used in tyre treads prior to the recent trend towards solution rubbers. Tan δ was measured with a dynamic testing system (MTS Systems Corporation) under conditions (40 Hz, 60°C, 5% strain) that relate to rolling resistance. With the exception of NR, the rubbers fall on a broad band which clearly shows that, as lower tan δ values are

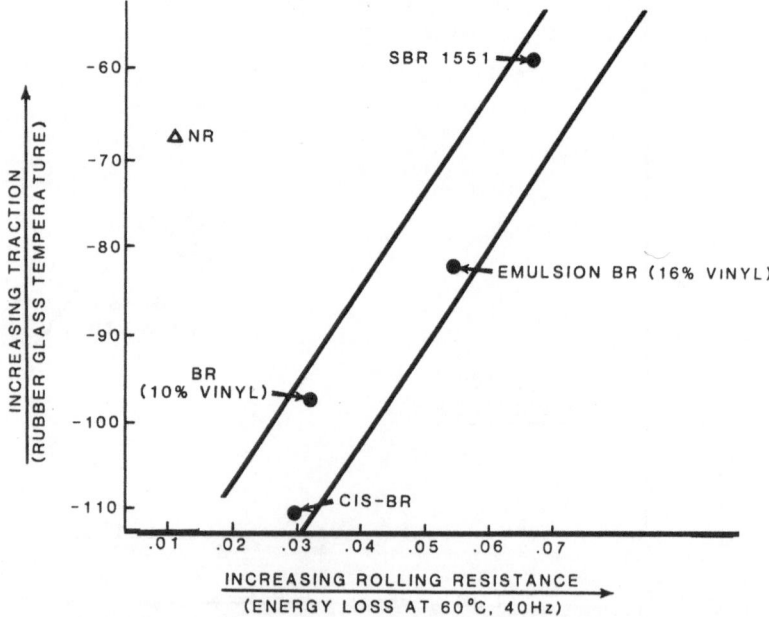

Fig. 26. Conventional tread rubber behaviour. (Reproduced with permission from ref. 62, by courtesy of Plenum Publishing Corporation, New York.)

approached for lower rolling resistance, there is a sacrifice to reduced wet traction, i.e. lower glass transition. Natural rubber (NR), which has a very low tan δ, does not have the level of wet traction attainable in some of the synthetic rubbers.

Having examined the relationship between T_g and tan δ for conventional tread rubbers, it is important to examine those structural features of SBRs that affect T_g and loss tangent, as shown in Table 10. The major features which influence T_g are the composition of the rubber and chain microstructure; macrostructure parameters have a lesser influence on T_g.[61] With respect to those variables influencing tan δ, the effects of molecular weight (free chain ends), MWD, crosslink density and molecular chain architecture are known to a greater extent than the influence of the chain structure aspects (emphasized in this Table by the question-mark entries). Very recently, the studies of Aggarwal et al. have defined the relationships between molecular structure of solution BRs and SBRs and their loss tangent behaviour.[60,62] Before we present the results of this study (Section 4.4), it is useful to review briefly the technological information concerned with the vinyl structure of solution polymers. The earlier work pointed to the unique dynamic properties of solution vinyl-BRs and the departure from their classical relationship to T_g.

TABLE 10

MOLECULAR FEATURES OF RUBBERS DERIVED FROM BUTADIENE AND STYRENE AND THEIR RELATIONSHIP TO T_g AND LOSS TANGENT[a]

Molecular structural features	T_g	Loss tangent (temp. $\gg T_g$ frequency—low)
Composition	✔	?
Chain microstructure (% vinyl, cis and trans)	✔	?
Monomer sequence distribution	✔	?
Molecular weight and molecular weight distribution	✔	✔
Crosslink density	✔	✔
Molecular chain architecture (linear or star)	✔	✔

[a] Data reproduced with permission from ref. 62, by courtesy of Plenum Publishing Corporation, New York.

4.3. Retrospective View of Vinyl-BR Properties

In the 1950s, the Phillips Petroleum Company and the Firestone Tire and Rubber Company started commercial production of polybutadienes by organolithium polymerization for use in tyres. These solution BRs, having low vinyl contents (8–10%), were used in blends with emulsion SBR in tyre treads for balancing traction and wear performance properties. In the early 1970s when styrene monomer was in short supply, developments from Phillips Petroleum Company[63] and EniChem[64] (formerly the International Synthetic Rubber Company) showed that vinyl-BRs with 50–55% vinyl content behaved like emulsion polymerized SBR in tyre tread formulations and exhibited very similar tread wear and wet skid resistance. Tread compounds containing 45%-vinyl polybutadiene showed lower heat build-up and better blow-out resistance than E-SBR and blends of E-SBR with cis-BR. EniChem introduced trial quantities of a medium-vinyl butadiene rubber (MVBR) under the name Intolene 50[65] in 1973.

Nordsiek and Kiepert also recognized the value of the vinyl structure for imparting non-typical dynamic properties.[66] Their work showed that under severe test conditions, vinyl-BRs exhibited appreciable improvements in resisting heat build-up and resisting thermal degradation, relative to both emulsion and solution SBRs when compared at equivalent T_g. Of special significance is the much greater temperature dependence of tan δ for vinyl-BRs. Loss tangent values in the region of the viscoelastic spectrum related to rolling resistance were shown to decrease in the order: E-SBR > S-SBR > vinyl-BR.

A more recent development in solution polymer technology concerns vinyl-BRs with special structures, reported by Nordsiek.[67] The disclosed rubbers, described as 'Integral Rubbers', have an arrangement of varying vinyl sequences in the form of multi-block sequences of vinyl units. The concept of 'Integral Rubbers' illustrated in the viscoelastic spectrum (Fig. 27) can be described as a summation of the viscoelastic responses of polymers having varying T_g. Tyre test data indicate that the rolling resistance of the Integral Rubber is similar to NR and its wet traction matches E-SBR 1712. Abrasion resistance was shown to be intermediate to NR and E-SBR 1500.

4.4. Influence of Styrene and Vinyl Contents on Wet Traction and Rolling Resistance

In a recent paper,[62] Aggarwal et al. studied the effect of changing styrene and pendant vinyl group content on T_g (wet traction) and

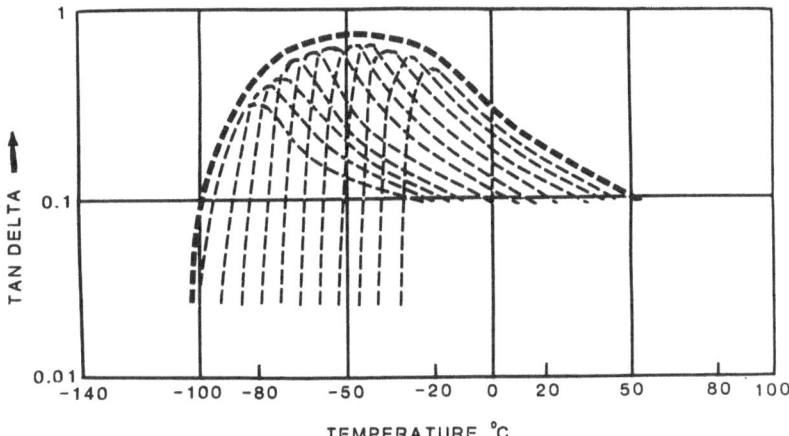

FIG. 27. Concept of Integral Rubbers.

tan δ. For a series of random solution SBRs at a constant percentage of vinyl structure (10% vinyl), the effects of styrene content on T_g and tan δ are shown in Fig. 28. As expected from the behaviour of emulsion SBRs, both T_g and tan δ increase linearly with increasing percentages by weights of styrene. Thus, like conventional tyre tread

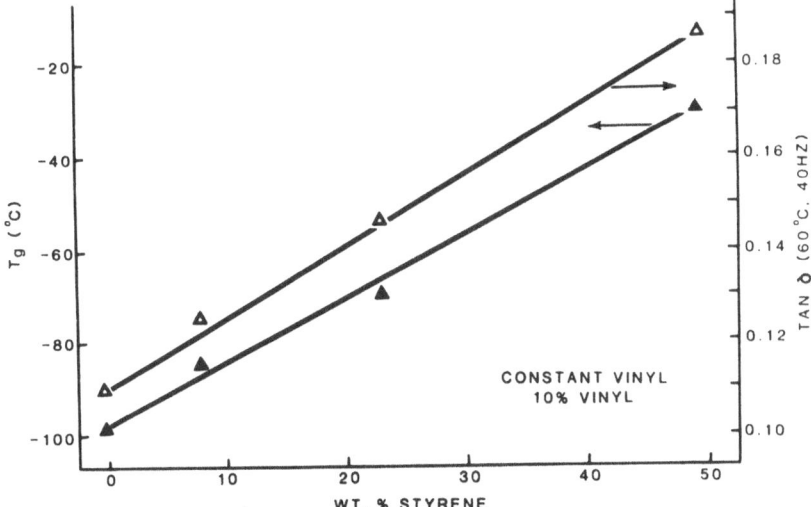

FIG. 28. Variation of T_g and tan δ with styrene content of solution SBRs at constant (10%) vinyl structure. (Reproduced with permission from ref. 62, by courtesy of Plenum Publishing Corporation, New York.)

rubbers, a trade-off of lower wet traction for lower rolling resistance is characteristic of low-vinyl solution SBRs.

For polybutadienes of varying vinyl contents at essentially equivalent molecular weight, the effect of vinyl structure on tan δ is unique, as shown in Fig. 29. Up to about the 50% level, tan δ is essentially independent of the proportion of vinyl structure. Thus, T_g of polybutadienes, or their wet traction, can be increased as the vinyl content is increased up to 50%, at the same time not adversely affecting tan δ or rolling resistance. At higher vinyl contents above 50%, tan δ increases sharply. Plots of tan δ versus vinyl content of vinyl-SBRs at a constant percentage of styrene (15 wt% styrene) have a similar shape to Fig. 29, indicating that the effect is not restricted to polybutadiene. This is a previously unrecognized characteristic feature of vinyl-substituted BRs and SBRs with important implications for obtaining an improved combination of high traction and low rolling resistance.

Although the vinyl structure of solution polymers is a dominating

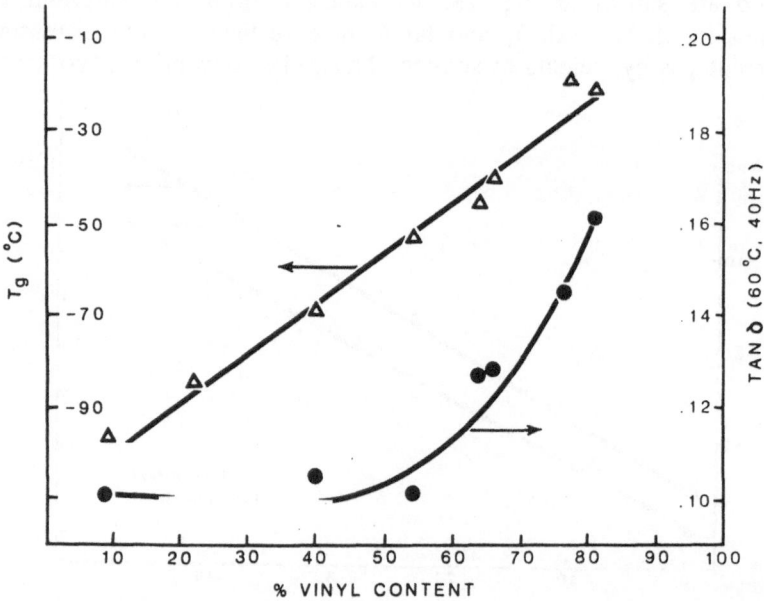

FIG. 29. Variation of T_g and tan δ with vinyl content of vinyl-BRs. (Reproduced with permission from ref. 62, by courtesy of Plenum Publishing Corporation, New York.)

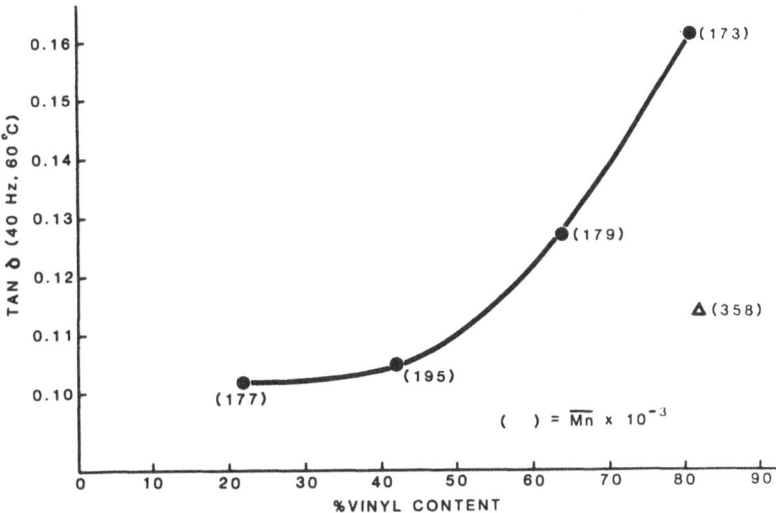

FIG. 30. Tan δ dependence of vinyl content at comparable molecular weights.

factor in determining the viscoelastic properties, molecular weight and MWD are also extremely important considerations, since the hysteresis of a vulcanizate increases with the number of free chain ends.[68,69] The molecular weight dependence of tan δ can be demonstrated by returning to the tan δ versus vinyl content relationship. Livigni reported[70] (Fig. 30) the same tan δ dependence on vinyl content for four vinyl-BRs all having similar high molecular weight values. However, the extent of increase in tan δ for HVBR is drastically reduced if molecular weight is increased from 173 000 to 358 000 as shown in the high vinyl region of the graph.

4.5. Blends of Solution Polymer with other Diene Rubbers

Although tread vulcanizates containing only a solution rubber have superior dynamic properties, their main deficiency is their poor ultimate properties, such as tear strength and tensile strength, which relate to wear. The need for improved abrasion resistance in solution rubber is indicated by a comparison of laboratory data (Table 11) for black filled vulcanizates of E-SBR, solution SBR (S-SBR) and MVBR at nearly equivalent T_g. For E-SBR, it is possible that a higher *trans*/*cis* ratio of the polybutadiene portion and a broader MWD may relate to its better abrasion resistance (Pico Index). The relative values

TABLE 11

RELATIVE TREAD PROPERTIES OF EMULSION AND SOLUTION SBR (BR) OF
SIMILAR GLASS TRANSITION TEMPERATURE

Polymer	Styrene (%)	Diene structure(%)		$T_g(°C)$ (onset)	Tread properties		
		Vinyl	trans		Rolling resistance index	Wet skid index	Pico index
E-SBR	14	18	72	−66	100	100	100
S-SBR	14	35	41	−63	71	102	65
MVBR	0	45	34	−64	67	102	76

of rolling resistance and wet skid indices are entirely consistent with
the polymer structure effects already discussed.

It is not surprising that blends of solution polymers with other diene
rubbers have generated considerable interest in order to meet simultaneously the processing and performance needs (Table 12) of tyre
tread rubbers. Blends of E-SBR/cis-BR developed previously for a
good balance of traction, measured by an instrumented portable skid
tester (IPST), and wear are simply too high in tan δ, measured by an
instrumented Yerzley oscillograph (IYO), as shown in Fig. 31. Natural
rubber alone gives very low rolling resistance, but its wet traction is
not favourable for passenger tyre treads. However, NR can provide
needed improvements in processability and hot tear resistance. Blends
of NR and E-SBR give lower rolling resistance than E-SBR/cis-BR
blends, but at a sacrifice in abrasion resistance. Blends of NR with
solution SBRs (and BRs) provide further improvements in both rolling
resistance and wet traction;[71] however, again abrasion resistance is

TABLE 12

PERFORMANCE PROPERTIES OF TREAD RUBBERS

(1) Low-energy losses/cycle (low hysteresis)
    (a) Low heat generation
    (b) Low rolling resistance
(2) Traction
    (a) Wet traction
    (b) Dry traction
(3) Durability
    (a) Wear

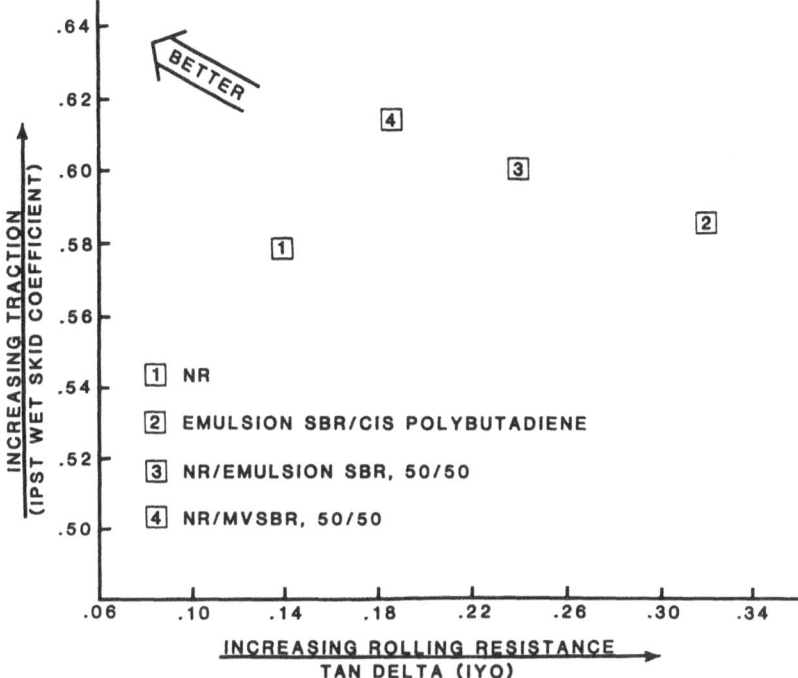

FIG. 31. Traction/rolling resistance for conventional polymers and a NR/MVSBR blend.

inferior to blends based on *cis*-BR. Blends containing *cis*-BR and vinyl-BRs appear to offer the fewest trade-offs in the combination of functional properties; however, adjustments in compound formulations may be needed for acceptable hot tear strength and processability.

4.6. Solution Rubbers Developed for Improved Abrasion Resistance, Rolling Resistance and Wet Traction

In order to describe the tyre tread performance capabilities of solution rubbers completely, we need to consider their wear resistance as well as rolling resistance and wet traction. Recent developments concerning abrasion resistance of synthetic rubbers have been announced by GenCorp and Bridgestone. Ogawa *et al* have disclosed MVSBRs, prepared with *n*-BuLi and a dual modifier system, with an increased *trans*-1,4-butadiene structure.[72] Limited increases in *trans*-1,4-content, to as high as 60%, are reported to provide improved wear resistance with excellent wet skid resistance and rolling resistance. In an adjoining patent reference, MVSBRs (medium-vinyl) having an increased $\bar{M}_w$ and bimodal MWD are claimed to provide improved resistance to uneven wear of tyre treads.[73]

A major development in solution SBRs was first disclosed at the International Rubber Conference in Kiev by R. A. Livigni et al.[58] and in subsequent publications.[24,74] As described in Section 3 of this chapter, these solution SBRs have a butadiene portion of trans-1,4-content 80–90%, vinyl content 2–4%, and a random comonomer sequence distribution. Additionally, the polymerization process, which uses a new catalyst of Mg–Al alkyls complexed with barium alkoxide salts, has all the distinguishing characteristics of organolithium-initiated polymerization (Tables 13 and 14).

By controlling the amount of trans-1,4-structure and the level of styrene, the crystalline melt temperature can be adjusted such that the new synthetic rubbers (HTSBR) are essentially amorphous at room temperature and have the capability to undergo strain-induced crystallization. This phenomenon also occurs with NR and, as a result, it will be shown that for uncured compounds HTSBRs exhibit both high green strength and tack strength (Section 4.7). As a consequence of its rather low T_g, this synthetic material has very good abrasion resistance, being superior to both general-purpose emulsion and solution SBRs (Table 15). Moreover, blends with only moderate amounts of

TABLE 13
SOLUTION SBR STRUCTURAL CONTROL

Using conventional anionic polymerization techniques

(1) Copolymer composition and sequence distribution.
(2) Ratio 1,4-(trans + cis)/1,2 (vinyl)polybutadiene microstructure.
(3) Molecular weight and molecular weight distribution.
(4) Extent and type of chain branching.

Using alkaline-earth-based polymerization catalyst developed at GenCorp

(In addition to those features listed above)
(5) Ratio of trans-1,4/cis-1,4-polybutadiene at low vinyl content: high-trans SBR (HTSBR).

TABLE 14
CHARACTERISTICS OF HTSBRs

(1) Random solution SBR of general-purpose type.
(2) Combination of high-*trans*-1,4 and low-vinyl content.
(3) 'Living polymerization' characteristics—controlled
    molecular weight, MWD and chain architecture.
(4) Variable degree of crystallinity, adjustable crystalline
    melt temperature.
(5) Good green strength, tack strength and processability.
(6) Low hysteresis, reduced flex crack growth rate
    (in blends with NR) and high abrasion index.

NR show excellent tack and green strength, reduced flex crack growth rate and very low heat build-up (Table 16).

Further developments in the technology of HTSBRs indicate a breakthrough to an improved combination of high skid and abrasion resistance together with low rolling resistance. Table 17 shows a comparison for tread compositions containing NR and an MVSBR (15% sytrene, 45% vinyl) in a 50/50 ratio, a 40/60 blend of *cis*-BR/MVBR, and a 30/70 blend of HTSBR/HVBR. Based on the nearly equivalent tan δ values, all of these compositions have comparable rolling resistance. However, both of the all-synthetic rubber compositions have an improved abrasion resistance (Pico Index). The blend of *cis*-BR/MVBR (45% vinyl) does show a slightly lower wet traction value; however, this property is improved at an

TABLE 15
COMPARISON OF TREAD VULCANIZATE PROPERTIES OF HTSBR WITH
OTHER SBRs

Physical properties[a]	HTSBR	Solution SBR	Emulsion SBR
M-300 (MPa)	12	11	12
Tensile strength (MPa)	24	17	22
Elongation at break (%)	490	425	510
Hardness, Shore A	68	70	69
Wet skid resistance index	88	83	100
Pico abrasion index	132	120	100

[a] Compound formulation (phr): polymer (100), N351 carbon black (50), naphthenic oil (10), ZnO (3), stearic acid (2), antioxidant (1·8), TBBS (1·4), sulphur (1·8).

TABLE 16

COMPARISON OF PHYSICAL PROPERTIES OF HTSBR, AN HTSBR/NR BLEND
AND A cis-BR/NR BLEND

Property	BR/NR 55/45	HTSBR/NR 60/40	HTSBR 100%
Tack (MPa at 30 s)	0·21	0·46	0·29
Green strength (MPa)	0·33	1·25	2·0
Modulus at 100% (MPa)	1·3	1·5	1·3
Modulus at 300% (MPa)	5·0	6·3	5·0
Tensile strength (MPa)	16·6	17·0	16·4
Elongation at break (%)	710	610	750
Goodrich heat build-up (100°C), ΔT(°C)	33	22	28
Crack growth (%):			
Pierced, 100 000 flexes at 25°C	30	32	42
Pierced, 10 000 flexes at 100°C	65	48	69
Unpierced, 475 000 flexes at 100°C	←No crack initiation→		

accompanying sacrifice in the level of abrasion resistance. The blend of
HTSBR (7% styrene, 80% trans) with HVSBR (10% styrene, 60%
vinyl) clearly shows advantages in having a unique combination of
properties: low rolling resistance, high wet traction, high abrasion
resistance. The blends of HTSBR/HVBR also were found to have
better (or good) extrusion processability behaviour. Such a combination of desirable performance properties is shown by either a blend of
the homopolymers or a corresponding block copolymer of the AB-
type [HTSBR-b-HVSBR]. A variety of styrene–butadiene block
copolymers for improved tyre tread compositions have been disclosed
in the patent literature;[75–77] however, the majority of the patents do

TABLE 17

COMPARISON OF TREAD PROPERTIES OF BLENDS OF HTSBR/HVSBR WITH
BLENDS OF NR/MVSBR AND cis-BR/MVBR

Tread rubber composition		Properties		
Low-T_g polymer/high-T_g polymer	Blend ratio	tan δ IYO	μ-WET IPST	Abrasion resist. index
HTSBR/HVSBR	30/70	0·17	0·64	111
NR/MVSBR	50/50	0·17	0·60	87
cis-BR/MVBR	40/60	0·16	0·53	110

not recognize the use of solution SBRs as the only tread rubber component, and none of them describes a crystallizable SBR segment having high-*trans*-1,4 structure together with low vinly content.

4.7. Tyre Carcass Properties of Crystallizable Solution Rubbers

The properties of building tack and green strength are lacking in solution rubbers produced by conventional anionic polymerization techniques. In order to achieve the levels of tack and green strength that are needed in radial tyre carcass construction, blends with NR and/or the use of tackifying resins are often used. It would be advantageous to be able to obtain the properties of green strength and building tack that are characteristic of NR, in solution polymers.

First, it should be mentioned that a number of approaches have been employed to increase the green strength of synthetic rubbers, including a variety of chemical modification methods, precuring with electron beam radiation and the development of new synthetic rubbers with strain-crystallizing behaviour.[78] Examples of such polymers that exhibit strain crystallization and develop good green strength are high-*trans* SBRs,[74] *trans*-polypentenamers[79] and *trans*-butadiene–piperylene copolymers.[80] Factors such as the availability of low-cost monomers and the use of expensive catalysts have deterred the production of the latter two rubbers on a commercial scale. These problems are avoided in strain-crystallizing HTSBRs described in Section 3. They promise improvements in green strength as well as tack strength, as discussed below, and can be produced from readily available monomers with a catalyst system that is competitive economically with alkyllithiums.

The shape of the uniaxial stress–strain plot for an uncrosslinked rubber provides a useful indication of green strength (Fig. 32). Stress–strain curves of an HTSBR (15% styrene, 85% *trans*), NR and E-SBR 1500 are compared for unvulcanized compounds containing 45 phr HAF carbon black. HTSBR and NR both have high green strength as evidenced by the positive slope beyond strains of about 200% in contrast to the negative slope for the case of E-SBR. Thus, like NR, HTSBR exhibits outstanding green strength, whereas E-SBR is very poor in this property.

Building tack or tack strength, i.e. the force required to separate two uncured polymer surfaces after they have been brought into contact, is another important property desired in a tyre carcass rubber. Of course, the upper limit of tack strength is a rubber's green strength.

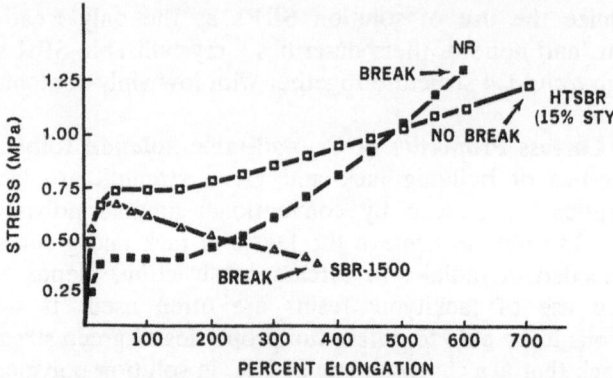

FIG. 32. Comparison of green strength of HTSBR, NR and E-SBR. (Reproduced with permission from ref. 24, by courtesy of the American Chemical Society, Washington, DC.)

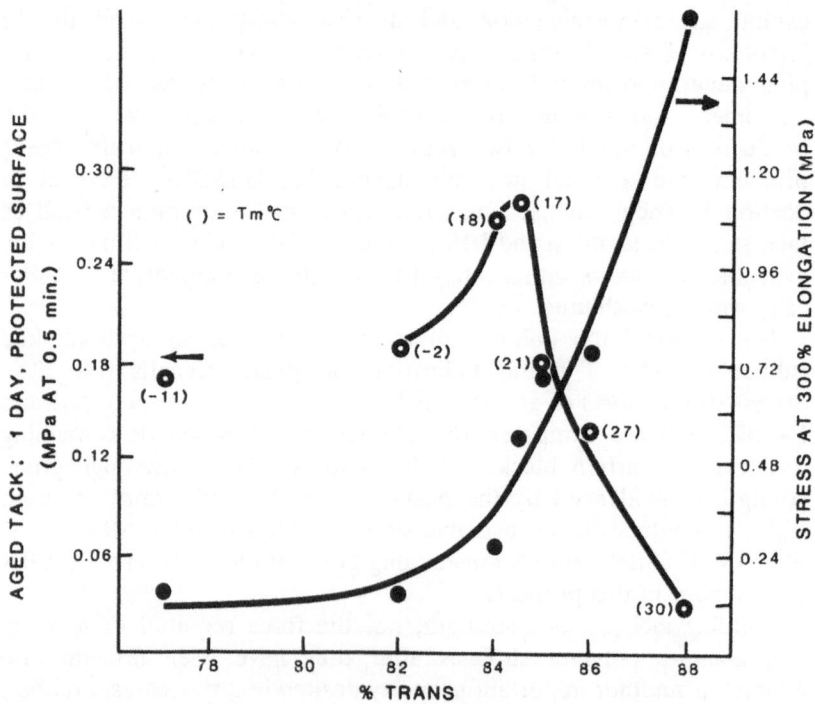

FIG. 33. Variation of tack and green strength with *trans*-1,4-content of HTSBRs (15% styrene).

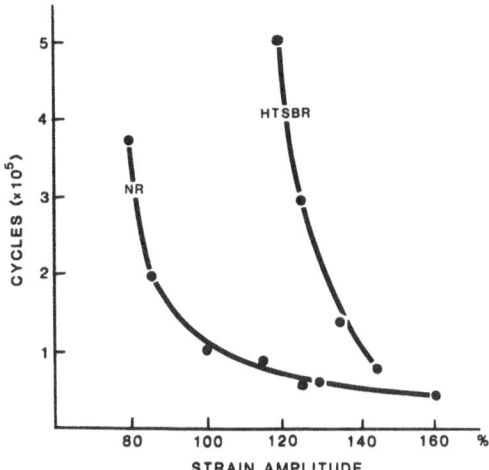

FIG. 34. Cycles to failure versus strain amplitude for vulcanizates of NR and HTSBR with no crack initiation.

In order not to hinder segmental diffusion across the rubber compound interface, which requires viscous flow under low stress conditions, nascent crystallinity must not be too high under in-use conditions. The T_m of the rubber must be near or slightly above the test temperature. The two curves in Fig. 33 demonstrate that the rubber formulations based on HTSBR of about 84–85% *trans* structure provide an optimum in tack strength and a high green strength.

We have discussed above the uncured properties of a synthetic rubber which are similar to those of NR. A major difference between the rubbers is that the synthetic SBR has better thermal oxidative resistance than NR. Under oxidative ageing conditions, 1,4-polybutadiene structure tends to crosslink to a greater extent relative to undergoing chain scission. The reverse is the case for NR (*cis*-1,4-polyisoprene structure). The greater resistance to oxidative degradation of HTSBR vulcanizates is indicated by a comparison of plots of flex life (cycles to failure) of HTSBR and NR versus strain amplitude, shown in Fig. 34.

5. PROSPECTS FOR SOLUTION RUBBERS

Innovation in new polymers and modifications of existing ones will be needed to meet future demands for tyre applications. Significant contributions may be expected not only in microstructure and macro-structure, and how they relate to the fundamental properties of rubbers, but also in their role in the morphology of polymeric materials. Innovations will be forthcoming in the area of specific interactions of polymers with fillers, extenders, curatives and the use of polymeric compatibilizers for rubber blends. Rubbers prepared with anionic polymerization catalysts such as organolithiums and alkaline earths have an exciting potential for making further improvements in the performance and processing properties of tyre components, as well as in other applications.

REFERENCES

General References

BLACKLEY, D. C., *Synthetic Rubbers: Their Chemistry and Technology*, Applied Science Publishers, London, 1983.

BRYDSON, J. A., *Rubber Chemistry*, Applied Science Publishers, London, 1978.

BRYDSON, J. A., 'Styrene–butadiene rubber,' in *Developments in Rubber Technology—2*, ed. A. Whelan and K. S. Lee, Applied Science Publishers, London, 1981.

HALASA, A. F., SCHULZ, D. N., TATE, D. P. and MOCHEL, V. D., 'Organolithium catalysis of olefin and diene polymerization', in *Advances in Organometallic Chemistry*, ed. F. G. A. Stone and R. West, Vol. 18, Academic Press, New York, 1980.

MCGRATH, J. E., ed., *Anionic Polymerization, Kinetics, Mechanisms and Synthesis*, American Chemical Society Symposium Series No. 166, ACS, Washington, DC, 1981.

MORTON, M., *Anionic Polymerization: Principles and Practice*, Academic Press, New York, 1983.

YOUNG, R. N., QUIRK, R. P. and FETTERS, L. J., *Adv. Polym. Sci.*, **56**, 1984, 1.

Literature Cited

1. ZIEGLER, K., DERSCH, F. and WILLTHAN, H., *Ann.*, **511**, 1934, 13.
2. ZIEGLER, K. and JAKOB, L., *Ann.*, **511**, 1934, 45.
3. ZIEGLER, K., JAKOB, L., WILLTHAN, H. and WENZ, A., *Ann.*, **511**, 1934, 64.
4. MORTON, M., BOSTICK, E. E. and LIVIGNI, R. A., *Rubber and Plastics Age*, **42** (4), 1961, 397.
5. SZWARC, M., LEVY, M. and MILKOVICH, R., *J. Amer. Chem. Soc.*, **78**, 1956, 2656.

6. PETERSON, C. J. and FRENSDORFF, H. F., *Angew. Chem., English Edn,* **11,** 1978, 161.
7. HALASA, A. F., SCHULZ, D. N., TATE, D. P. and MOCHEL, V. D., *Advances in Organometallic Chemistry,* ed. F. G. A. Stone and R. West, Vol. 18, Academic Press, New York, 1980, p. 55.
8. MORTON, M. and FETTERS, L. J., *Rubber Chem. Technol.,* **44** (3), 1975, 359.
9. HALASA, A. F., LOHR, D. F. and HALL, J. E., *J. Polym. Sci., Polym. Chem. Ed.,* **19,** 1981, 1357.
10. URANECK, C. A., *J. Polym. Sci., Part A-1,* **9,** 1971, 2273.
11. ANTKOWIAK, T. A., OBERSTER, A. E., HALASA, A. F. and TATE, D. P., *J. Polym. Sci., Part A-1,* **10,** 1972, 1319.
12. BYWATER, S., MACKERRON, D. H. and WORSFOLD, D. J., *J. Polym. Sci., Polym. Chem. Ed.,* **23,** 1985, 1997.
13. MILNER, R., YOUNG, R. N. and LUXTON, A. R., *Polymer,* **26,** 1985, 1265.
14. LUXTON, A. R., BURRAGE, M. E., QUACK, G. and FETTERS, L. J., *Polymer,* **22,** 1981, 382.
15. QUACK, G. and FETTERS, L. J., *Macromolecules,* **11,** 1978, 369.
16. MCELROY, B. J. and MERKLEY, J. H., US Patent 3 678 121 assigned to Lithium Corp., 1972.
17. ENGEL, E. F., *Gummi. Asbest. Kunst.,* **26,** 1973, 5.
18. GLAGUE, A. D. H., VAN BROEKHOVEN, J. A. M. and BLAAUN, L. P., *Macromolecules,* **7,** 1974, 348.
19. KRAUS, G., CHILDERS, C. W. and GRUVER, J. T., *J. Appl. Polym. Sci.,* **11,** 1967, 1581.
20. RAKOVA, G. V. and KOROTKOV, A. A., *Dokl. Akad. Nauk SSSR,* **119,** 1958, 982.
21. KOROTKOV, A. A. and CHESNOKOVA, N. N., *Vysokomolekul. Soedin.,* **2,** 1960, 365.
22. KOROTKOV, A. A. and RAKOVA, G. V., *Vysokomolekul. Soedin.,* **3,** 1961, 1482.
23. HSIEH, H. L. and WOFFORD, C. F., *J. Polym. Sci., Part A-1,* **7,** 1969, 461.
24. HARGIS, I. G., LIVIGNI, R. A. and AGGARWAL, S. L., in *Elastomers and Rubber Elasticity,* ed. J. E. Mark and J. Lal, Symposium Series No. 193, American Chemical Society, Washington, DC, 1982.
25. KOLTHOFF, I. M., LEE, T. S. and CARR, C. W., *J. Polym. Sci.,* **1,** 1946, 429.
26. MOCHEL, V. D. and JOHNSON, B. L., *Rubber Chem. Technol.,* **43,** 1970, 1138.
27. TANAKA, Y., SATO, H., SAITO, K. and MIYASHITA, K., *Rubber Chem. Technol.,* **54,** 1981, 685.
28. WERSTLER, D. D., *Rubber Chem. Technol.,* **53** (5), 1980, 1114.
29. BYWATER, S., *Adv. Polymer Sci.,* **4,** 1965, 66.
30. ZELINSKI, R. R. and HSIEH, H. L., US Patent 3 280 084 assigned to Phillips Petroleum Co., 1966.
31. BI, L. K. and FETTERS, L. J., *Macromolecules,* **9,** 1976, 732.

54   I. G. HARGIS, R. A. LIVIGNI AND S. L. AGGARWAL

32. VITUS, F. J., HARGIS, I. G., LIVIGNI, R. A. and AGGARWAL, S. L., US Patent 4 497 748 assigned to GenCorp Inc., 1985.
33. MORRISON, R. C. and KAMIENSKI, A. W., US Patent 3 725 368 assigned to Lithium Corporation of America, 1973.
34. FARRAR, R. C., US Patent 3 787 510 assigned to Phillips Petroleum Co., 1974.
35. VITUS, F. J., HARGIS, I. G., LIVIGNI, R. A. and AGGARWAL, S. L., US Patent 4 553 578 assigned to GenCorp Inc., 1985.
36. VITUS, F. J., HARGIS, I. G., LIVIGNI, R. A. and AGGARWAL, S. L., US Patent 4 426 495 assigned to GenCorp Inc., 1984.
37. TREPKA, W. J., US Patent 3 393 182 assigned to Phillips Petroleum Co., 1968.
38. URANECK, C. A. and SHORT, J. N., *J. Appl. Polym. Sci.*, **14**, 1970, 1421.
39. NEUMANN, W. P., *The Organic Chemistry of Tin*, Interscience Publishers (John Wiley and Sons), London, 1970, p. 34.
40. YOSHIMURA, N., OKUYAMA, M. and YAMAGISHI, K., *ACS Rubber Division 122nd Meeting, Chicago, Oct. 1982*, Paper 31.
41. SAITO, Y., *Kautsch. Gummi Kunst.*, **39**, 1986, 30.
42. FUJIMAKI, T., OGAWA, M., YAMAGUCHI, S., TOMITA, S. and OKUYAMA, M., Bridgestone Corp., *International Rubber Conference (IRC), Kyoto, Japan, 1985*.
43. OSHIMA, N., SAKAKIBARA, M. and TSUTSUMI, F., Japan Synthetic Rubber Co., *Proc. Internat. Institute of Synthetic Rubber Producers (IISRP), Buenos Aires, Argentina, 1985*.
44. TUNG, L. H. and BEYER, D. E., US Patent 4 172 100 assigned to Dow Chemical Co., 1979.
45. BROSKE, A. P., HUANG, T. L., HOOVER, J. M., ALLEN, R. D. and McGRATH, J. E., *Polymer Preprints*, **25** (2), 1984, 85.
46. BYWATER, S., 'Anionic polymerization', in *Encyclopedia of Polymer Science and Engineering*, Vol. 2, John Wiley and Sons, New York, 1985, p. 38.
47. QUIRK, R. P., IGNATZ-HOOVER, F. and CHEN, Wei-Chih, *Polymer Preprints*, **27** (1), 1986, 188.
48. SCHULZ, D. N., SANDA, J. C. and WILLOUGHBY, B. G., in *Anionic Polymerization, Kinetics, Mechanisms and Synthesis*, ed. J. E. McGrath, Symposium Series No. 106, American Chemical Society, Washington, DC, 1981, Chapter 27.
49. SCHULZ, D. N., HALASA, A. F. and OBERSTER, A. E., *J. Polym. Sci., Part A-1*, **12**, 1974, 153.
50. TATE, D. L. and GRAVES, D. F., European Patent Application 177695 assigned to Firestone Tire & Rubber Co., 1986.
51. NOGUCHI, K., YOSHIOKA, A., KOMURO, K. and UEDA, A., 'Structure and properties of newly developed chemically modified high vinyl polybutadiene and solution polymerized styrene–butadiene rubbers', *ACS Rubber Division 129th Meeting May, 1986*, Paper No. 36.
52. CAMPBELL, D. R. and WARNER, W. C., *Anal. Chem.*, **37**, 1965, 276.
53. KELEN, T. and TÜDOS, F., *J. Macromol. Sci.—Chem.*, **A9** (1), 1975, 1.

54. NATTA, G. and CORRADINI, P., *Rubber Chem. Technol.*, **33**, 1960, 703.
55. FLORY, P. J., *Principles of Polymer Chemistry*, Cornell University Press, 1953, p. 569.
56. FOX, T. G., *Bull. Amer. Phys. Soc.*, **2** (2), 1956, 123.
57. STEIN, R. S., *Rheology*, ed. F. R. Eirich, Vol. V, Academic Press, New York and London, 1969, p. 279.
58. LIVIGNI, R. A., HARGIS, I. G. and AGGARWAL, S. L., 'Structure and properties of rubbers for tyres and new developments for crystallizing butadiene rubbers,' Paper presented at *International Rubber Conference (IRC), Kiev, USSR, 1978.*
59. MANDELKERN, L., *Crystallization of Polymers*, McGraw Hill, New York, 1964, p. 264.
60. AGGARWAL, S. L., HARGIS, I. G., LIVIGNI, R. A., FABRIS, H. J. and MARKER, L. F., *Polymer Preprints*, **26** (2) 1985, 3.
61. BRYDSON, J. A., *Rubber Chemistry*, Applied Science Publishers, London, 1978.
62. AGGARWAL, S. L., HARGIS, I. G., LIVIGNI, R. A., FABRIS, H. J. and MARKER, L. F., 'Structure and Properties of Tire Rubbers Prepared by Anionic Polymerization,' in *Advances in Elastomers and Rubber Elasticity*, Plenum Press, New York, 1986.
63. RAILSBACK, H. E. and STUMPE, N. A., *Rubber Age*, **107** (12), 1975, 27.
64. DUCK, E. W., *Europ. Rubber J.*, **155** (11), 1973, 25.
65. ANON., *Chem. Eng. News*, **52** (25), 1974, 13.
66. NORDSIEK, K. H. and KIEPERT, K. M., *Kautsch. Gummi Kunst.*, **35**, 1982, 371.
67. NORDSIEK, K. H., *Kautsch. Gummi Kunst.*, **38** (3), 1985, 178.
68. MANCKE, R. G. and FERRY, J. P., *Trans. Soc. Rheol.*, **12**, 1968, 335.
69. TAKAO, H. and IMAI, A., *International Rubber Conference (IRC), Kyoto, 1985.*
70. LIVIGNI, R. A., *International Rubber Conference (IRC), Gothenburg, Sweden, 1986.*
71. TATSUO, F., YAMAGUCHI, S., YAMADA, T. and TOMITA, S., EP 48 619 assigned to Bridgestone Tire Co. Ltd, 1982.
72. OGAWA, M. and IKEGAMI, M., US Patent 4 387 756 assigned to Bridgestone Tire Co. Ltd, 1983.
73. OGAWA, M. and IKEGAMI, M., US Patent 4 387 757 assigned to Bridgestone Tire Co. Ltd, 1983.
74. FABRIS, H. J., HARGIS, I. G., LIVIGNI, R. A. and AGGARWAL, S. L., 'New synthetic rubber with high green strength: high *trans* SBR,' *ACS Rubber Division, 124th Meeting, Houston, Texas, 1983*, Paper No. 89.
75. FUJIMAKI, T., YAMADA, T. and TOMITA, S., US Patent 4 396 743 assigned to Bridgestone Tire Co. Ltd, 1983.
76. TAKEUCHI, Y., SAKAKIBARA, M., TSUTSUMI, F., TAKASHIMA, A. and HATTORI, I., US Patent 4 433 109 assigned to Japan Synthetic Rubber Co. Ltd, 1984.

77. HATTORI, Y., IKEMATU, T., IBARAGI, T. and HONDA, M., US Patent
    4 413 098 assigned to Asahi Chemical Co., 1983.
78. BUCKLER, E. J., SHACKLETON, J. and WALKER, J., *Elastomeric*, **114,**
    1982, 17.
79. HAAS, F. and THREISEN, D., *Kautsch. Gummi Kunst.*, **23,** 1970, 502.
80. LAURETTI, E., SANTARELLI, G., CANDIDO, A. and GARGANI, L., *Proc.
    Internat. Rubber Conference, Venice, 1979*, p. 322.

Chapter 2

ADVANCES IN NITRILE RUBBER (NBR)

P. W. MILNER

Compagnie Française Goodyear, Les Ulis, France

1. INTRODUCTION

Nitrile rubber (NBR) is the accepted polymer for oil, solvent, fuel and grease resistance. Over 300 grades are now commercially available from which the rubber industry can choose polymers that will satisfy the most stringent demands of modern rubber mixing and processing methods. The expertise in compound formulation design built up over the 40 years since NBR was first introduced has also made it possible to improve the final properties of NBR compounds in order to meet new service conditions.

In those areas, however, where the service environment has become even more severe, the polymer chemist has responded with NBR modifications that have markedly improved the mechanical performance of the elastomers without affecting oil and fuel resistance.

Taken as a group, nitrile rubbers are still the best cost/performance elastomers available for oil and fuel resistance and remain the standard choice for these types of applications.

2. CHARACTERISTICS OF NBR

Since the last review of butadiene–acrylonitrile copolymers (NBRs) appeared in this series in 1981 several new trends in the development

57

of NBR have become apparent and some ideas have now reached the
stage of commercialization.

2.1. General Properties

This continual search to improve the properties of a polymer which
was first produced on a commercial scale in 1937 only serves to
highlight how important the basic characteristics of nitrile polymers
are to the rubber-using sectors of industry.

These characteristics have been well documented in the past by W.
Hoffman in his comprehensive review of 1963[1] and earlier by W. L.
Semon;[2] to these references can be added the continual output of
technical literature supplied by the commercial producers of NBR.

The general properties of NBR have been summarized by Bertram[3]
and are shown in Table 1.

The purpose of this chapter is to bring together the latest developments that have occurred in order to improve some of the above-
mentioned properties. In doing this the objective has been to
concentrate on those developments that have produced commercially
available polymers that are of immediate use to the industry rather
than to cover in depth all the associated research ideas that have
appeared in the literature.

TABLE 1
GENERAL PROPERTIES OF NBR

+	Oil, fuel and grease resistance
+	Good processing characteristics
+	Variety of curing systems
+	Good hot air resistance
	Long term: 90°C
	40 days: 120°C
	3 days: 150°C
+	Low permanent set
+	Good abrasion resistance
+	Low gas permeability
−	Moderate to good low temperature flexibility
−	Moderate ozone resistance (except NBR/PVC)
−	Moderate tack
+	Compatibility with polar thermoplastics
	(e.g. PVC, phenolics)

2.2. Polymer Modifications

As for most olefin-containing elastomers, the presence of the residual double bond in the nitrile rubber backbone enables conventional sulphur crosslinking to occur, but this carries with it the inherent weakness of susceptibility to oxygen and ozone attack which imposes limits on the final operating conditions.

A considerable amount of work has been expended on extending the working range of nitrile rubber, so that its invaluable properties of oil and fluid resistance can meet the more severe operating conditions now encountered. Typical of these changing conditions is the use of higher engine temperatures in the motor vehicle industry and the requirements of the oil industry for elastomeric parts needed to help extract the more difficult oil deposits from offshore locations.

Three such polymer modifications designed to improve the service characteristics of NBR are now commercially available. These are the saturated (or hydrogenated) grades, the antioxidant bound types and the carboxylated grades.

In addition to these developments, work has been carried out to improve the standard range of nitrile rubbers by designing polymers with better colour and reduced mould-fouling characteristics, and an attempt to rationalize the range and thereby simplify polymer choice has also occurred.

The modification of NBR by PVC has also been the subject of development work and the influence of carboxylation on the NBR in the blend has been studied.

Lastly, thermoplastic NBRs have now made their commercial debuts after extended product evaluations, and will certainly find an industry 'niche'.

The developments will now be dealt with in more detail.

3. HIGHLY SATURATED AND HYDROGENATED NITRILE RUBBERS

The replacement of the oxidatively sensitive unsaturation in butadiene–acrylonitrile rubbers renders the polymer more resistant to heat ageing both in air and when in contact with or immersed in fluids. This change in the configuration of the molecular chain can be brought about in several ways.

3.1. Obtaining Saturation

A hydrogenation of the already-polymerized nitrile polymer can be done by subjecting solutions of the polymer to either low- or high-pressure action of hydrogen in the presence of various catalysts. Complexes of rhodium, ruthenium, cobalt and iridium have all been used with various degrees of success.[4-8] In some cases the hydrogenation is incomplete and a certain degree of unsaturation remains which allows the polymer to be conventionally vulcanized, whereas in other cases complete saturation of the polymer is obtained and this requires the use of peroxides for crosslinking. The degree of hydrogenation achieved appears to be a function of the solvent used for dissolution; chlorine-substituted aromatic solvents completely hydrogenate the double bonds whereas solvents such as benzene, toluene and cyclo-hexanone lead to only selective hydrogenation.

An alternative method of obtaining a fully saturated NBR-type polymer is to copolymerize alternately ethylene and acrylonitrile which gives a polymer with oil resistance comparable with conventional NBR coupled with markedly improved oxygen and ozone resistance.[9]

3.2. Properties of Saturated NBR

A range of three highly saturated NBRs (HSNBRs) has been described by Hashimoto et al.[10] in which a certain degree of unsatura-tion still remains, thus allowing crosslinking to take place with conventional sulphur accelerator systems. Table 2 compares one of these polymers with a standard NBR for physical properties before and after exposure to the typical sour gas and fluid conditions found in the oil well industry.

Two other commercially available polymers in which the saturation is complete have been described by Mirza and Thormer.[11] Two types are proposed which contain 17% and 19% lateral nitrile groups corresponding to 34% and 38% conventional acrylonitrile content. These polymers behave much like other rubbers and can be mixed on either two-roll mills or in internal mixers. A typical mix formula is shown in Table 3.

To obtain the optimum ageing properties the use of a styrenated diphenylamine antioxidant in conjunction with the zinc salt of mercap-tobenzimidazole is recommended. Peroxide curing is needed and the higher activation temperature types are favoured. In addition a cyanurate-type co-agent is required to give optimum crosslinking for minimum compression set characteristics.

TABLE 2
HSNBR versus standard NBR

	A	B
Formulation		
HSNBR (45% ACN, 80 ML4)	100·0	—
NBR (41% ACN, 75 ML4)	—	100·0
Zinc 2-mercaptotoluimidazole (ZMTI)	1·5	1·5
Substituted diphenylamine (Naugard 445)	1·5	1·5
Stearic acid	0·5	0·5
Zinc oxide	5·0	5·0
N330 Black (HAF)	50·0	60·0
Sulphur	0·5	1·5
TMTD	1·5	—
MBT	0·5	—
TMTM	—	0·2
TETD	1·0	—
MBTS	—	1·5
DOP	—	5·0
Properties		
ML4 at 100°C	99	112
Rheometer at 160°C		
t_5 (min)	3·1	2·5
t_{90} (min)	20·0	13·4
Vulcanization time (min at 160°C)	20·0	20·0
Post cure (h at 150°C)	4·0	—
Hardness (° Shore A)	84	84
Tensile strength (MPa)	28·3	29·1
200% Modulus (MPa)	19·5	22·7
Elongation at break (%)	310	270
Ageing in sour gas phase (4·8% H_2S, 20% CO_2, 75·2% CH_4) (168 h at 150°C)		
Change in tensile strength (%)	−7	−84
Change in elongation at break (%)	−55	−98
Hardness change (° Shore A)	−7	−8
Ageing in sour liquid phase (95% diesel oil, 4% H_2O, 1% NACE amine B) (168 h at 150°C)		
Change in tensile strength (%)	−29	−87
Change in elongation at break (%)	−55	−92
Hardness change (° Shore A)	−9	−2

TABLE 3
HYDROGENATED NBR—TYPICAL FORMULA

Formulation (*phr*)	
HNBR	100·0
Stearic acid	1·0
Magnesium oxide	10·0
Zinc oxide	2·0
PE wax	1·0
Styrenated diphenylamine antioxidant (SPDA)	1·5
Zinc methylmercaptobenzimidazole (ZMMBI)	1·5
Black N-550 (FEF)	50·0
Triallyl cyanurate	2·5
1,3-Bis(*tert*-butylperoxyisopropyl)benzene	7·5
Compound Properties	
Density (g/cm^3)	1·22
Mooney viscosity ML 1 + 4/100°C	121·0
Mooney Scorch MS-t_s/140°C (min)	15·0
Rheometer at 170°C	
t_{10} (min)	2·5
t_{90} (min)	15·0

The physical properties of the formula shown in Table 3 are listed in Table 4. The resultant resistance to hot-air ageing is shown in Fig. 1.

The behaviour of these polymers when subjected to conditions similar to those encountered below ground in an oil well are compared with a conventional NBR and a fluorinated elastomer in Fig. 2. The

TABLE 4
PHYSICAL PROPERTIES OF HYDROGENATED NBR
VULCANIZED FOR 20 MIN AT 170°C

Tensile strength (MPa)	23
Elongation at break (%)	300
100% modulus (MPa)	6·1
200% modulus (MPa)	17·7
Hardness (° Shore A)	72
Rebound resilience (%)	40
Tear resistance, DIN 53507 (N/mm)	7
Abrasion loss, DIN 53516 (mm^3)	
60-grade emery paper	80

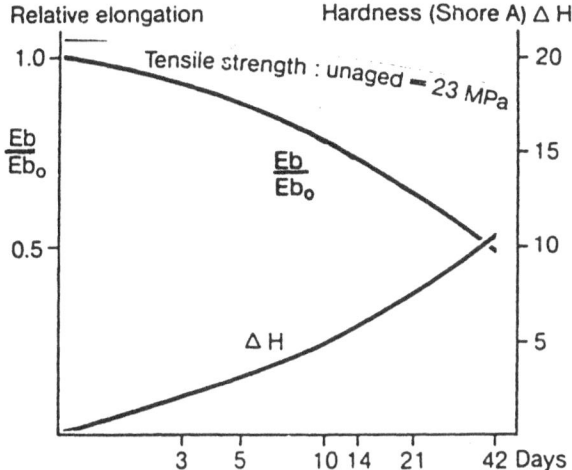

FIG. 1. Change in tensile strength, elongation at break and hardness after ageing at 150°C. (Reproduced with permission from ref. 11.)

composition of the crude oil reference fluid was (%wt):

ASTM oil No. 3	99·5
Diaminobenzene	0·25
N,N'-Di(1,4-dimethylpentyl)-p-phenylenediamine	0·25

The relative speed with which hydrogenated or highly saturated nitriles have appeared on the market commercially[12] has highlighted the need for polymers that fill the performance gap between standard NBRs and the high-performance fluorinated elastomers. This is an area of polymer and technology development that is in a rapid state of change and more candidates for this market, defined by the automobile and oil exploration industries, are sure to appear.

4. BOUND-ANTIOXIDANT NITRILE RUBBERS

The use of network-bound antioxidants to prevent losses due to extraction and volatilization has been the subject of considerable research work since the late 1960s.

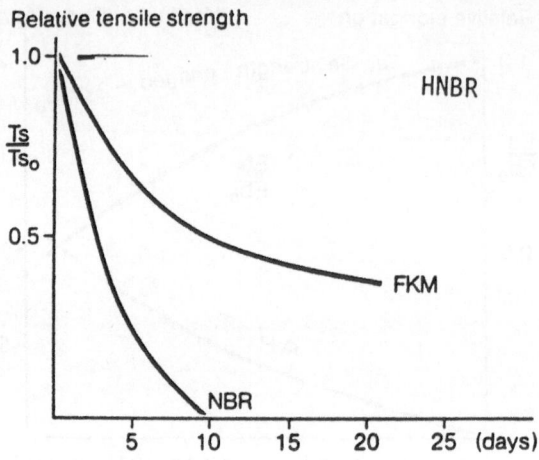

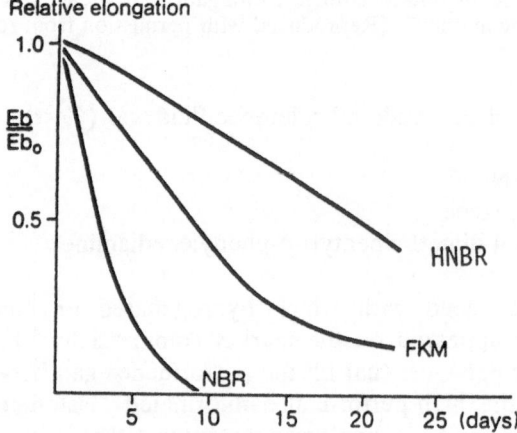

FIG. 2. Change in tensile strength and elongation at break after immersion in a crude oil reference fluid at 150°C in the presence of H$_2$S. (Reproduced with permission from ref. 11.)

4.1. Attaching the Antioxidant

Scott *et al.* have proposed several methods by which antioxidant molecules can be attached to the polymer chain without loss of activity. Nitrones and in particular *N*-phenylnitrones can be chemically bound to the chain during the vulcanization process[13-15] and they can

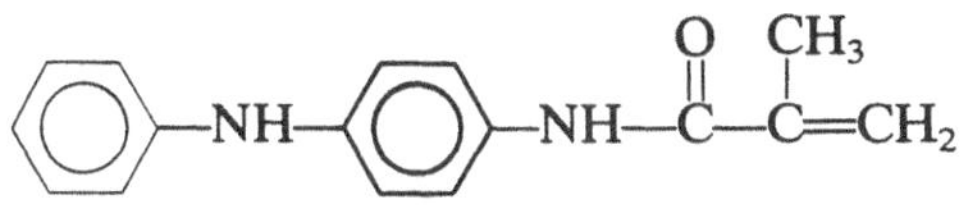

FIG. 3. N-(4-anilinophenyl)methacrylamide monomeric antioxidant.

also be grafted onto the vinyl groups of the NBR chain in the latex phase after polymerization.[16] The same, bound-antioxidant effect is achieved by grafting stabilizers onto the 4-methylene groups in the polymer chain and aromatic antioxidants which themselves contain 4-methylene groups can be grafted to polymer backbones.[17,18]

Mori and co-workers[19] have proposed antioxidant substituted triazine compounds, which act as combined vulcanizing agents and antioxidant systems which resist the leeching action of fluids during immersion, particularly in 'sour' gasoline. Another method involving the use of thiophenols and amine-substituted thio compounds has been described by Weinstein[20,21] where the attachment is achieved during or after the polymerization process.

The types of antioxidant that can be copolymerized into polymer chains has been reviewed.[22] The use of a vinyl-terminated antioxidant useful for copolymerization has been proposed by Kline and Miller[23] and their incorporation into SBR and NBR was described by Meyer *et al.*[24] The use of copolymerizable monomers containing acrylamine antioxidant substituents,[25] and in particular N-(4-anilinophenyl)methacrylamide[26] (Fig. 3), appears to be the most favoured.

4.2. Test Results

Nitrile rubber containing the antioxidant specified in Fig. 3 exhibits improved oxygen absorption resistance both before and after methanol extraction (Fig. 4). This property has been confirmed in final product application testing where the non-volatility and non-extractability of the stabilizer guarantees a long-term antioxidant effect.[27-30] A range of antioxidant-bound NBR polymers has now been commercialized[31] and a comparison of the typical physical property differences of the medium–low acrylonitrile content polymer when compared with a conventional NBR is shown in Table 5.

To obtain the maximum results from these polymers certain compounding rules have been established. Vulcanization should be

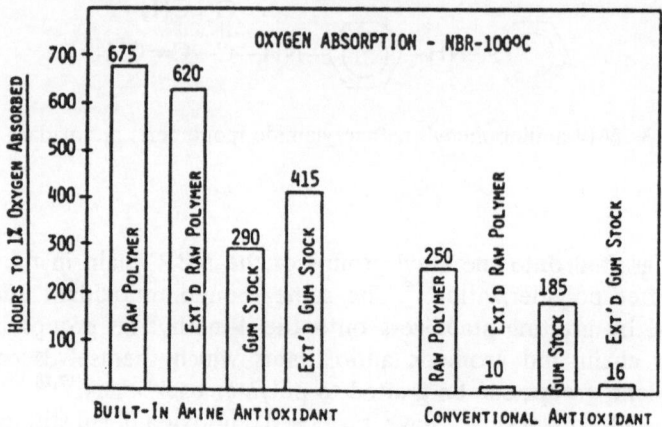

FIG. 4. Oxygen absorption of NBR before and after extraction with methanol.
Built-in antioxidant: 1·25 pts N-(4-anilinophenyl)methacrylamide. Added
antioxidant: octylated diphenylamine. (Reproduced with permission from ref.
24.)

TABLE 5

PHYSICAL PROPERTY DIFFERENCES BETWEEN AO-BOUND AND
CONVENTIONAL NBR[a]

	AO-bound NBR	Conventional NBR
Formulation (phr)		
Chemigum HR 765	100·0	—
Conventional nitrile (27% ACN)	—	100·0
Zinc oxide	3·0	3·0
Stearic acid	1·0	1·0
Hi-Sil 215	40·0	40·0
Silane A-189	0·4	0·4
Magnesium oxide	10·0	10·0
MT Black	5·0	5·0
Hercoflex 600	10·0	10·0
Flectol H	—	3·0
TE-80	2·0	2·0
Sulphur MC	0·2	0·2
TMTD	2·0	2·0
Morfax	1·0	1·0

TABLE 5—*contd.*

	AO-bound NBR	Conventional NBR
Properties		
Rheometer cure at 162°C		
TM (lb in)	59·2	59·8
TL (lb in)	16·5	13·7
t_2 (min)	2·5	2·4
t_{90} (min)	13·3	10·5
Press cure for 15 min at 162°C		
Tensile strength (MPa)	14·83	16·89
Elongation at break (%)	610	755
300% Modulus (MPa)	5·52	3·80
Hardness (° Shore A)	53	56
Compression set B(%)		
70 h at 121°C	14·2	34·7
70 h at 150°C	38·9	62·3
ASTM No. 1 oil for 168 h at 150°C		
Tensile strength (MPa)	6·55	5·86
Elongation at break (%)	400	310
Hardness (° Shore A)	52	58
Vol. swell (%)	−5·3	−5·0
ASTM No. 3 oil for 168 h at 150°C		
Tensile strength (MPa)	2·07	4·83
Elongation at break (%)	320	435
Hardness (° Shore A)	25	34
Vol. swell (%)	37·3	25·6
Sequence ageing: 70 h in ASTM No. 1 oil at 150°C + 70 h in air oven at 150°C		
Tensile strength (MPa)	10·34	3·45
Elongation at break (%)	515	70
Hardness (° Shore A)	54	72
Sequence ageing: 70 h in ASTM No. 3 oil at 150°C + 70 h in air oven at 150°C		
Tensile Strength (MPa)	9·45	4·83
Elongation at Break (%)	360	50
Hardness (° Shore A)	71	79
Solenoid brittleness (°C)	−40	−34
Gehman low-temp. torsion (°C)		
T_2	−18·0	−20·0
T_5	−26·0	−26·0
T_{10}	−29·0	−28·0
T_{100}	−40·0	−37·0

[a] AO, antioxidant-bound.

carried out with low sulphur, sulphur donor or peroxides, and filler systems should be based on blends of carbon black (preferably thermal grades) and fine particle mineral types. Initial work favoured cadmium oxide as an activator, but this has now been replaced for toxicity reasons by magnesium oxide.

4.3. Applications
The main applications for these types of polymer are in automotive fuel hoses, pump seals, power steering units and transmission gaskets.

5. CARBOXYLATED NBR

The introduction of carboxylation into a conventional butadiene–acrylonitrile elastomer was first reported in 1946[32] and the use of acrylic or methacrylic acid as a termonomer yields polymers which have enhanced tensile strength, modulus, tear resistance and hardness coupled with exceptionally high abrasion resistance.[33]

The presence of the polar carboxyl groups also adds to the inherent oil and solvent resistance imparted by the acrylonitrile groups and the overall swelling resistance to fluids is improved.[33]

5.1. Crosslinking
The use of the carboxylic group reactivity as a possible source of crosslinking was not initially recognized as being of any practical significance until the mid-1950s[34,35] when the use of polyvalent metal oxides and diamines as crosslinking agents was reported. The key to the usefulness of carboxylated nitriles lies in the use of these pendant carboxyl groups for crosslinking. The affinity of this group to react with a metal oxide yields ionic crosslinks which give the high physical properties already cited. These enhanced properties are attributed to the presence of hard ionic clusters dispersed throughout an amorphous rubbery matrix.[36] These ionic clusters can be considered as acting as a reinforcing filler. Unfortunately this type of crosslink shows a lower thermal stability than a sulphur type resulting in unfavourable compression set at elevated temperatures.[37] Covalent sulphur crosslinking can also take place and, using a low-sulphur or sulphur donor system, acceptable compression set resistance can be obtained.

It is modern practice to use both systems in formulating carboxylated nitrile rubbers in order to obtain the optimum balance of properties. Ideally, it is preferable that the two types of crosslinking take place at the same rate but unfortunately in practice the metal–carboxylate reaction takes place more rapidly and leads to scorching.[33]

5.2. Scorch Control

The scorching problem is aggravated by moisture or humidity which accelerates the zinc oxide–carboxylic acid reaction rate, and can only be overcome by either surface treatment of the zinc oxide with zinc sulphide or zinc phosphate[38] or by the use of more costly zinc peroxide, preferably in masterbatch form.[39,40] The influence that these variations has on scorch time is illustrated in Table 6.[41]

TABLE 6

XNBR　GUM COMPOUNDS SHOWING EFFECTS OF CURATIVE MASTERBATCHES ON SCORCH TIME

Formulation (pbw)			
Carboxylated NBR	95		
Sulphur MC	1·5		
Stearic acid	0·5		
TE-80	2		
Hi-Sil	5		
TMTM	0·5		
Metal oxide⎫(A, B or C)[a] NBR ⎭	10		
	114.5		

	A	*B*	*C*
Properties			
Compound viscosity (ML-4 at 100°C)	45	45	46
Mooney scorch at 125°C (min.)	11	17	25
Optimum cure time at 166°C (min.)	17	21	23
Modulus at 100% elong. (MPa)	2	2	2
Modulus at 300% elong. (MPa)	5	4·5	4·5
Tensile strength (MPa)	36	32	35
Elongation at break (%)	550	550	560

[a] A, 50:50 NBS grade zinc oxide and 34 ACN, 35 Mooney NBR. B, 50% active, curative masterbatch of treated zinc oxide with inert filler, mechanically dispersed in a low-viscosity, medium-ACN nitrile rubber. C, 50:50 latex-dispersed, curative masterbatch of zinc peroxide in a medium-ACN nitrile rubber.

One disadvantage of these external scorch control systems is that they also have a retarding effect on optimum cure time which tends to negate any move towards faster cure cycles.

Another approach has been to incorporate at the polymerization stage oligomerized fatty acids that enhance the scorch safety without affecting the cure rate.[42-44] This polymer can use either zinc oxide or zinc peroxide, but zinc oxide is preferred for cost reasons. Proper precautions still have to be taken against the influence of moisture and a propionic acid-treated zinc oxide has proved to be effective.[45] This is shown in Table 7.

5.3. Applications

Carboxylated nitrile rubbers are now increasingly seen as versatile, easy-to-process abrasion-resistant elastomers which can be used in seals, valve linings, military and industrial footwear, hoses, pump components and conveyor belts, and perhaps more significantly in all aspects of oilfield applications.[47]

TABLE 7

SCORCH RESISTANCE OF INTERNALLY RETARDED XNBR (A) AND STANDARD XNBR (B)[46]

	A	B
Formulation (pbw)		
Polymer	100·00	100·00
GPF Black	40·00	40·00
Stearic acid	2·00	2·00
Dibutyl phthalate	5·00	5·00
Styrenated diphenylamine	1·00	1·00
Sulphur MC	1·50	1·50
TMTM	1·30	1·30
TMTD	0·20	0·20
Treated zinc oxide	5·00	5·00
	156·00	156·00
Properties		
Mooney scorch, small rotor, 121°C		
Minimum torque	17·00	45·00
t_s (min)	36·2	3·2

6. THERMOPLASTIC NITRILE RUBBERS

Since the introduction of the first thermoplastic elastomers (TPEs) in the 1960s the importance of TPEs has grown considerably due to the commercial advantages that they offer in having the properties of vulcanized rubber without the associated costs of the classical vulcanization process.

6.1. Two-Phase Structure

Thermoplastic elastomers are usually two-phase in nature, having a resinous thermoplastic phase which contributes the processing ease of a plastic and an elastomeric or rubber phase which contributes the elastic behaviour typical (and commercially exploitable) of vulcanized rubber. These two-phase mixtures can be prepared either by blending two polymers, or by preparing block polymers. In this latter case the elastomer phase is not crosslinked in any way, but derives its strength from the thermoplastic block segments that agglomerate during cooling after the processing stage. This can be considered as a type of crosslinking as the thermoplastic block segments are connected to the rubber segments by primary chemical bonds.[48]

By contrast the simple blending of two polymers can only yield a technically adequate product if some interaction occurs between the plastic and elastomeric domains. Polymers based on the dynamic vulcanization principles first outlined by Gessler[49] and extended by Coran and Patel[50,51] have successfully overcome this phase morphology problem.

6.2. Commercially Available TPEs

Abdou-Sabet et al.[52] have described a range of oil-resistant thermoplastic elastomers that are now commercially available. These polymers are prepared by the dynamic vulcanization of blends of nitrile rubber and a polyolefin resin. In order to overcome the gross incompatibility of these two polymers a compatibilizing mechanism needs to be introduced and it is suggested that the production of a block copolymer from the two incompatible polymers would be one way of achieving this.[53] As is common with other thermoplastic elastomers the grades are differentiated by hardness levels. The typical physical properties of the grades are shown in Table 8. Physical property changes and volume swell after immersion in comparison with standard NBRs are listed in Tables 9 and 10 and Fig. 5.[52] The test compounds used for the standard NBRs are shown in Table 11.

TABLE 8

PHYSICAL   PROPERTIES   OF   DIFFERENT   HARDNESS   GRADES   OF
THERMOPLASTIC NBRs[a]

Property	701–80	701–87	703–40
Hardness, (° Shore)	80A	87A	40D
Specific gravity	1·09	1·07	1·05
100% Modulus (MPa)	5·4	6·9	10·3
Tensile strength (MPa)	11·0	14·1	19·3
Elongation at break (%)	310	380	470
Tear strength (KN/m)	48	58	69
Tension set (%)	15	21	31
Compression set at 100°C for 22 h (%)	33	39	48
ASTM No. 3 oil volume swell at			
125°C for 70 h (%)	10	12	15
Brittle point (°C)	−40	−40	−36

[a] Data reproduced with permission from ref. 52.

TABLE 9

RETAINED PHYSICAL PROPERTIES AFTER IMMERSION IN ASTM OIL NO.3 AND
REFERENCE FUEL C[a]

Property		Standard nitrile	
	NBR–TPV	29% ACN	38% ACN
ASTM No. 3 oil (125°C/70 h)			
Change in hardness (° Shore A)	−5	−8	−9
100% Modulus (% retained)	97	89	89
Ultimate tensile strength (% retained)	70	97	97
Ultimate elongation at break (% retained)	54	100	88
Fuel C (23°C/70 h)			
Change in hardness (° Shore A)	+4	−4	−5
100% Modulus (% retained)	112	96	82
Ultimate tensile strength (% retained)	110	74	85
Ultimate elongation at break (% retained)	111	74	83

[a] Data reproduced with permission from ref. 52.

TABLE 10

VOLUME SWELL (%) OF A THERMOPLASTIC NBR COMPARED WITH STANDARD NBRs[a]

Fluid	Test time (h)	Test temp. (°C)	NBR–TPV 701–80	Nitrile	
				29% ACN	38% ACN
ASTM No. 3 oil	70	125	10	8·5	5·0
ASTM No. 2 oil	70	125	2	7	1·5
Gasohol[b]	70	23	12	20	22
Reference fuel A	70	23	3·5	6	5
Reference fuel C	70	23	18	25	24
Diesel fuel No. 2	70	23	12	7	3·5
Automatic transmission fluid	168	100	0	1·5	1
Brake fluid	168	100	23	25	33
Skydrol 500	168	100	87	80	86
Pydraul 312 C	168	100	74	55	52

[a] Data reproduced with permission from ref. 52.
[b] 90% Unleaded petrol + 10% ethanol.

TABLE 11

TEST COMPOUND FORMULATIONS FOR STANDARD NBRs[a]

Formulation (pbw)		
NBR (29% ACN)	100	—
NBR (38% ACN)	—	100
ZnO	5·0	5·0
Aminox	2·0	1·0
ZMTI	2·0	2·0
Maglite K	—	1·0
Stearic acid	1·0	1·0
Carbon black	120	110
Paraplex G-25	10·0	10·0
Sulphur	0·3	0·2
TMTD	4·0	1·5
CBS	3·0	—
Sulfasan R	1·5	1·5
Cure Characteristics, 160°C		
M_{HL} (lb in)	52	44
Cure time (min)	20	20

[a] Data reproduced with permission from ref. 52.

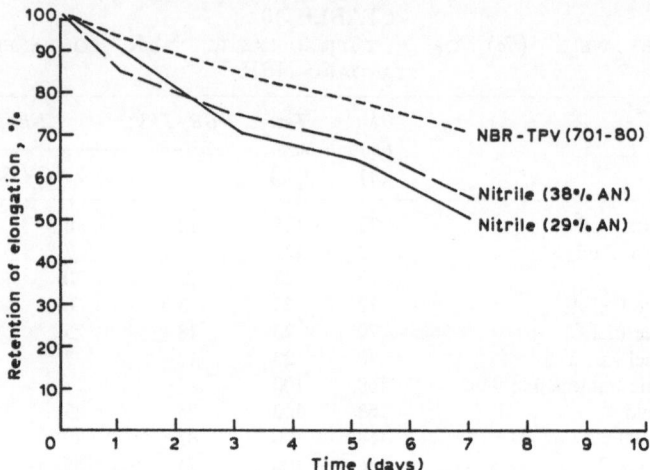

FIG. 5. Percentage retention of original elongation at break after hot-air ageing at 125°C. (Reproduced with permission from ref. 52.)

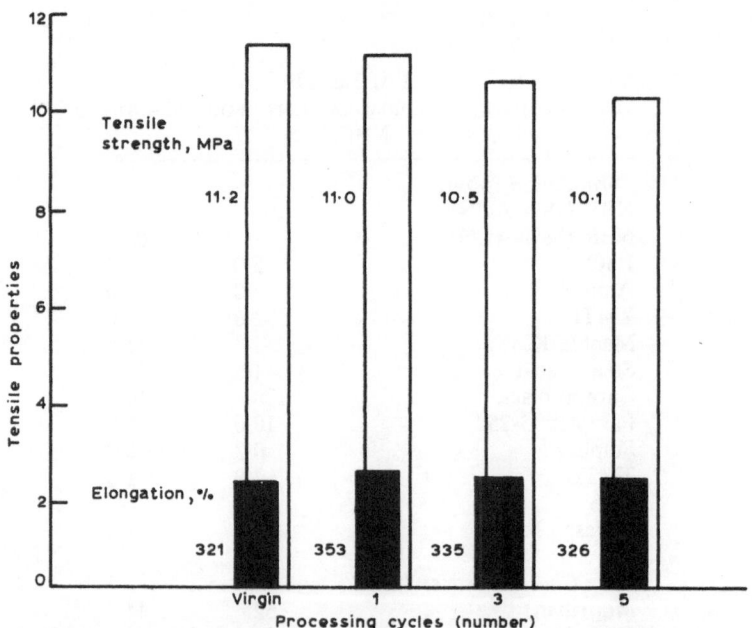

FIG. 6. Effect of recycling on physical properties of thermoplastic NBR–TPV Grade 701–80. (Reproduced with permission from ref. 52.)

TABLE 12
COMPRESSION SET (%) OF THERMOPLASTIC NBRs
VERSUS STANDARD TYPE[a]

Polymer	22 h at 20°C	22 h at 100°C
703–40	19·0	48·0
701–87	18·0	39·0
701–80	18·0	33·0
NBR (29% ACN)	5·0	27·5
NBR (38% ACN)	6·4	34·0

[a] Data reproduced with permission from ref. 52.

These polymers also show very little change in properties on recycling, as is shown in Fig. 6.

6.3. Applications of TPE

These polymers are aimed at the automotive, domestic appliance, fluid power and industrial hose industries, and although still not performing as well as conventional NBR on compression set (see Table 12), they could also find use in certain seals and gasketing applications.

Without doubt this is one of the most interesting developments that has taken place in recent years and will surely be followed by further types from other manufacturers.

7. NBR/PVC BLENDS

NBR can and has been used to modify a variety of plastics, but the most widely used blend is that with PVC. With this plastic it is possible to have two distinct types of blend ratios, firstly that where the PVC predominates and the resultant blend is processed as a thermoplastic and secondly that where the NBR predominates and the blend is subsequently processed as a vulcanizable elastomer.

This second type of blend ratio was initially developed as a means of improving the ozone resistance of nitrile rubber,[54] where it was used mainly as a competitor for polychloroprene rubber. The amount of PVC present in the blend can vary up to 50 parts, but to have any measurable improvement in weathering properties, at least 25 parts of PVC must be present in the blend. Most commercial producers of

nitrile rubbers have a range of NBR/PVC blends available, some of
which are mechanically prepared, though most are manufactured from
latex blends.[55]

7.1. Mechanical Blend Preparation
In the preparation of mechanical blends of PVC and NBR there are
two very important conditions which must always be satisfied if the
optimum properties are to be obtained. They are:

(a)   good dispersion or blending of the one polymer in the other;
(b)   satisfactory fluxing of the rubber into the PVC.

The blending of two solid polymers is best achieved when they have
similar Mooney viscosities. If the viscosities are widely different the
hard polymer tends to form discrete hard particles which float around
in the softer matrix formed by the other polymer. Consequently an
unhomogeneous mixture is formed which gives poor and inconsistent
results.

Therefore the mill or internal-mixer blending of solid PVC and
NBR polymers requires special treatment and careful control.

Mechanical blends of the solid polymers must also be fluxed at high
temperatures (150–160°C) to cause the complete homogeneity of the
two polymers and to obtain the optimum properties from the blend.[56]
Too low a temperature will result in underfluxed material with less
than optimum properties. Too high a temperature may result in
decomposition of the PVC, and possible gelation of the rubber.

The accurate control of the fluxing conditions is the critical factor in
obtaining acceptable homogenization and thereby the full advantages
of the blend.[57] This comment is true for the latex preblend types which
also need to be fluxed prior to use.

7.2. Comparison with Carboxylated NBR
The use of carboxylated NBR (XNBR) in blends with PVC has also
been studied[58,59] and the resultant blends exhibit increased tensile
strength and modulus with improved abrasion resistance. The main
drawback to the use of XNBR in such blends has been its tendency to
crosslink during the fluxing stage of blend preparation; this has now
been solved and proprietary blends can be obtained.[58]

A comparison of the property improvements that can be obtained
are shown in Tables 13 to 15.

TABLE 13

COMPARISON OF XNBR/PVC AND NBR/PVC FLUXED IN SIMPLE TEST RECIPES

Property	Rubber/PVC ratio					
	100:0		85:15		70:30	
	NBR	XNBR	NBR	XNBR	NBR	XNBR
Viscosity (ML1 + 4 min at 100°C)	67	103	61	93	58	83
Mooney scorch, t_s at 125°C (min)	25	16·5	25	15·2	25	13·7
Cure time at 166°C (min)	8	10	10	12	10	10
Specific gravity	1·14	1·14	1·17	1·18	1·20	1·20
Hardness (° Shore A)	63	76	69	80	73	82
Modulus at 100% elong. (MPa)	2·0	5·1	3·1	6·5	4·6	9·4
Modulus at 300% elong. (MPa)	10·1	20·6	10·5	19·2	12·0	20·6
Tensile strength (MPa)	22·3	22·3	20·2	21·8	17·3	21·6
Elongation at break (%)	600	350	600	360	530	340
NBS abrasion (%)	157	536	287	587	441	1032

7.3. PVC Modification

The use of NBR as a modifier for PVC has been well documented;[60] the improvements in the properties of a plasticized PVC compound are now well established[61] and specific grades of NBRs, produced in powder form for ease of dry blending, are now commercially available.[62] This modification of PVC could be considered to be outside

TABLE 14

FLUXED BLEND MASTERBATCHES [a] (pbw)

NBR (ACN 34%, ML: 50)	85	—	70	—
XNBR	—	85	—	70
PVC (medium MW)	15	15	30	30
Plasticizer	5	5	10	10
Stabilizer	2	2	2	2
Lubricants	1·75	1·75	2·5	2·5
Antioxidant	1·2	1·2	1·2	1·2
Mooney viscosity				
(ML1 + 4 min at 100°C)	46	57	46	55
(ML1 + 4 min at 125°C)	28	36	29	35

[a] Each stock was fluxed for 2 min at 155–160°C in an internal mixer, then sheeted off a warm (approx. 60°C) mill.

TABLE 15
BASE COMPOUNDS[a] (pbw)

Fluxed blend masterbatch	100
Sulphur	1·25
Stearic acid	1
N-550 (FEF) carbon black	40
TMTM	0·5
(50:50 NBR:ZnO) Masterbatch	10

[a] Compounds were mixed in a 4-minute, upside-down cycle; Model B, Banbury; 77 rpm; 80°C at start; cold water on full. The curative masterbatch and accelerators were added on the sheet-off mill.

the scope of this chapter, but with the increasing interest in thermo-plastic elastomers certain of these blends could be considered as possible candidates.[63,64] Developments are continuing in this field and a range of thermoplastic PVC/NBR polymers has recently been announced[65] but it is not yet possible to give an accurate technical assessment of their value.

8. DEVELOPMENTS IN STANDARD GRADES OF NITRILE RUBBER

The handbook of the International Institute of Synthetic Rubber Producers (IISRP)[62] lists over 300 standard grades of butadiene–acrylonitrile polymers which are mainly differentiated by acrylonitrile content. The acrylonitrile content particularly influences the potential fluid resistance and low-temperature flexibility.

8.1. NBR Classification
Polymers exist with acrylonitrile contents varying from 16 to 52% but these can be grouped into seven main categories as shown in Table 16.

This spectrum of polymers is further augmented by the fact that each grade is available at several raw-polymer viscosity levels.

The availability of such a range of polymers has been the result of constant requests from industry for polymers that can meet the ever-varying end-user product specifications.

TABLE 16

CLASSIFICATION OF NBRs
BY ACRYLONITRILE CONTENT

Grade	ACN (%)
Low	<23
Medium low	28
Medium	33
Medium high	38
High	41
Very high	45
Ultra high	>50

TABLE 17

PHYSICAL PROPERTIES OF 'RATIONALIZED' NBRs

Raw Polymer		
Acrylonitrile content (%)	30	40
Mooney viscosity, ML4	55	55
Compounded		
Modulus at 300% (MPa)	7·2	10·7
Tensile strenth (MPa)	16·9	18·9
Elongation at break (%)	660	570
Hardness (° Shore A)	58	65
LTF[a] (Gehman) ASTM D1053 T100(°C)	−31·0	−26·5
Volume Change		
ASTM oil No. 1—70 hrs at 125°C (%)	−9·2	−11·0
ASTM oil No. 3—70 hrs at 125°C (%)	+5·5	−3·1
ASTM fuel B—70 hrs at 23°C (%)	+24·0	+18·0
ASTM fuel C—70 hrs at 23°C (%)	+52·0	+28·0
Test Recipe		
Polymer	100·0	
Zinc oxide	5·0	
Stearic acid	1·0	
Antioxidant	1·5	
SRF Black	65·0	
DOP	15·0	
MBTS	1·5	
Sulphur	1·5	

[a] Low temperature flexibility.

8.2. Rationalization of Grades

In an attempt to rationalize the standard range, however, polymers have recently been introduced that are claimed[66] to combine at the two acrylonitrile levels of 30% and 40% the needs of low-temperature flexibility and oil resistance of all grades from medium low (28%) to high (41%) acrylonitrile content.

Some of the typical properties of two of these polymers are shown in Table 17.

This introduction of a simplified range is a first attempt to rationalize the existing range of NBRs, which is long awaited, but it is too early yet to judge how these polymers will be technically accepted and what their long-term effect will be on the development of standard grades.

9. MISCELLANEOUS DEVELOPMENTS

9.1. Powdered Grades

NBRs in powdered form are available in a number of viscosity values and acrylonitrile contents, and can offer the user considerable savings in processing costs by eliminating bale cutting, reducing mixing cycles with lower power consumption, and giving better dispersion and lower mixing temperatures with associated lower heat histories for the finished compounded stocks. The advantages come from using dry blending techniques to obtain the dispersion followed with short mixing cycles in internal mixers or mixer/extruders which generate the shear necessary for maximum physical properties.[67-69] Table 18 gives an indication of the economies that accrue.[70,71]

The main point of discussion now is whether there is a real saving in capital investment and whether the advantages justify the premium charged for the raw material. The question of capital investment would appear, at first sight, to be one of simple arithmetic but, in practice, rubber mixing machinery lasts so long that the cost of new plant often has to be compared with the 'written-down' value of existing equipment. Added to this, the present overcapacity of mixing facilities in the rubber industry has mitigated against any factory-scale development work and drastically reduced the speed with which this technology has been accepted.

The main outlet now for powdered NBR is for the modification of

TABLE 18
ECONOMIC STUDIES OF TOTAL PRODUCTION COSTS FOR EXTRUSION

Bale rubber	Relative cost (%)	Powder rubber	Relative cost (%)
(1) Classical two-stage mixing (2 × internal mixer) + strip cutting	100	(1) Turbo rapid mixer, compactor, mill + stock surcharge	62
(2) Two-stage mixing (1 × internal mixer, final mix process on mill) + strip cutting	88	(2) Turbo rapid mixer, compactor + mixing extruder (fixed costs) + stock surcharge	57
(3) Single-stage mixing + strip cutting	62		

plastics, particularly PVC.[60] These thermoplastic blends are being used to produce articles that had previously been the domain of vulcanized general-purpose polymers (e.g. shoe soles, work boots, extruded gaskets and sealing strips, calendered sheet, etc.).

9.2. Liquid Grades

Liquid forms of NBR have been available for over 30 years and serve primarily as non-volatile and non-extractable plasticizers for solid NBR compounds. The advantage of their use is that they are co-vulcanizable with the base NBR polymer employed, but their disadvantage is that they are viscous and extremely difficult to weigh and handle.

The use of liquid NBRs with reactive end groups has been reported[72] but their use as castable prepolymers to produce industrial goods has not progressed as the incorporation of reinforcing fillers, to give the required physical properties, has proved to be too difficult.

However, progress has been made in the use of carboxy-terminated NBRs to modify phenolics and epoxies for adhesive work. A comprehensive review of this area to which the reader is referred, has been carried out by Pocius.[73]

9.3. Grafted Polymers

In addition to the polymer modifications already discussed at the beginning of the chapter, an acrylate graft polymer has also been

TABLE 19

EFFECTS OF GRAFTED AND HOMOPOLYMERIZED 2-ETHYLHEXYL ACRYLATE
ON THE PROPERTIES OF NITRILE RUBBER

Plasticizer	Clash–Berg temperature (°C)		Volume swell (%) in ASTM No. 3 oil— 70 h at 125°C
	Original	Oil-aged	
None	−27	−30	20
Grafted acrylate[a]	−36	−39	22
Homopolymerized acrylate[a]	−31	−41	33

[a] At 20 phr

described by Eldred.[74] A 2-ethylhexyl acrylate monomer was grafted onto nitrile rubber during the normal cure cycle of the elastomer and led to a significant improvement in low-temperature flexibility. The plasticization was permanent and twice as effective as an externally added, polymeric acrylic plasticizer (see Table 19).

This is an area of development that might answer the needs of the other major service condition requirement of lower service temperature without loss of oil resistance.

10. FUTURE PROSPECTS

The changes in final product properties and the environment in which they will have to function will be the determining factors in the future development of nitrile rubbers. During the last decade there has been a constant and, in the most part, successful attempt by polymer chemists to improve the basic acrylonitrile–butadiene molecule to meet the challenge of the changing working conditions of parts designed for the automotive and oil industries. The carboxylation, hydrogenation and bound antioxidant modifications of NBR have all increased the mechanical effectiveness of the polymer without detracting from its prime property of oil and solvent resistance.

The most significant change that could occur in the usage pattern of NBRs would be an increasing use of the thermoplastic alternatives, where the production cost reductions possible in the moulding process would be difficult to resist. Even though some final application conditions have warranted a move to more exotic polymers, the various NBRs now commercially available still exhibit a cost performance potential that makes these polymers the standard materials for oil-, fuel- and solvent-resistant rubber parts in the foreseeable future.

ACKNOWLEDGEMENT

The author wishes to thank The Goodyear Tire & Rubber Co. for permission to publish this paper.

REFERENCES

1. HOFFMAN, W., *Rubber Chem. Technol.*, **36**, 1963, 1.
2. SEMON, W. L., in *Synthetic Rubber*, ed. G. S. Whitby, J. Wiley & Sons, New York, 1954, Chapter 23.
3. BERTRAM, H. H., in *Developments in Rubber Technology—2*, ed. K. S. Lee and A. Whelan, Elsevier Applied Science Publishers, London, 1981.
4. FINCH, A. M., Shell Oil Company, US Patent 3 700 637, 24 October, 1972.
5. WEINSTEIN, A. H., *Rubber Chem. Technol.*, **57**, 1984, 203.
6. BAYER AG, German patent DT 2 539 132, 1977.
7. OPPELT, D., SCHUSTER, H., THORMER, H. J. and BRADEN, R., British Patent 1 558 491, 3 January 1980.
8. NIPPON ZEON., Japanese Patent 78/39 744, 1978.
9. USCHOLD, R. E. and FINLAY, J. B., *Appl. Polym. Symp.*, **25**, 1974, 205.
10. HASHIMO'O, K., WATANABE, N. and YOSHIOKA, A., *Rubber World*, **190**(2), 1984, 32.
11. MIRZA, J. and THORMER, H. J., "Therban". Un nouvel elastomère spécial", *Caoutchoucs et Plastiques*, (651), May 1985, 71.
12. EARLY, D. D., *Elastomerics*, **117**(11), Nov. 1985, 26.
13. SCOTT, G., *Europ. Polym. J. Suppl.*, 1969, 189.
14. SCOTT, G., *Macromol. Chem.*, **8**, 1973, 319.
15. SCOTT, G. and SMITH, K. V., *Rubber Chem. Technol.* **52**, 1979, 949.
16. SCOTT, G., *International Rubber Conference, Brighton*, May 1977.
17. AJIBOYO, O. and SCOTT, G., *Polym. Degrad. Stab.*, **4**, 1982, 397.
18. AJIBOYO, O. and SCOTT, G., *Polym. Degrad. Stab.*, **4**, 1982, 415.
19. MORI, K., TUNEISHI, M. and NAKAMURA, Y., *Int. Polym. Sci. Technol.*, (2), 1985, 56.
20. WEINSTEIN, A. H., *Rubber Chem. Technol.*, **50**, 1977, 641.
21. WEINSTEIN, A. H., *Rubber Chem. Technol.*, **50**, 1977, 650.
22. *Antioxidants—Recent Developments*, Noyes Data Corporation, USA, 1979, 232.
23. KLINE, R. H. and MILLER, J. P., *Rubber Chem. Technol.*, **46**, 1973, 96.
24. MEYER, G. E., KAVCHOC, R. W. and NAPLES, F. J., *Rubber Chem. Technol.*, **46**, 1973, 106.

25. KLINE, R. H., US Patent 3 658 769, 25 April 1972.
26. HORVATH, J. W., PURDON, J. R., MEYER, G. E. and NAPLES, F. J., *Appl. Polym. Symp.*, **25**, 1974, 184.
27. HORVATH, J. W. and BUSH, J. L., *Proc. SAI Meeting*, Detroit, September 1978.
28. HORVATH, J. W., *Proc. ACS Rubber Division Meeting*, Atlanta, March 1979, Paper 23.
29. HORVATH, J. W., *Proc. SAE Meeting*, Detroit, June 1979.
30. HORVATH, J. W., *Elastomerics*, **111**(8), 1979, 19.
31. *Chemigum HR*, Technical Booklet, Goodyear Chemical Division, Akron, OH, April 1981.
32. SEMON, W. L. and GOODRICH, B. F., US Patent 2 395 017, 19 Feb. 1946.
33. BRYANT, C. L., *J. Inst. Rubber Ind.*, **4**, 1970, 202.
34. ROSAHL, D. *et al.*, Bayer AG, DE 955901, 12 Dec. 1956.
35. BROWN, H. P., *Rubber Chem. Technol.*, **30**, 1957, 1347.
36. TOBOLSKY, A. V., LYONS, P. F. and HATA, N., *Macromolecules*, **1**(6), 1968, 84.
37. CHAKRABORTY, S. K., BHOMICK, A. K. and DE, S. K., *J. Appl. Polym. Sci.*, **26**, 1981, 4011.
38. HALLENBECK, V. L., *Rubber Chem. Technol.*, **46**, 1973, 78.
39. WEIR, R. J. and BURKLEY, R. C., 'Carboxylated nitrile rubber: its properties and applications', New York Rubber Group, 1 Oct. 1981.
40. GUSEV, Yu. K. *et al.*, *Int. Polym. Sci. Technol.*, **12**(5), 1985, T20–T22.
41. WEIR, R. J. and GUNTHER, W. D., *Europ. Rubber J.*, Feb. 1978, 20.
42. GRIMM, D. C., Goodyear Tire & Rubber Co., US Patent 4 435 535, 6 March 1984.
43. GRIMM, D. C., Goodyear Tire & Rubber Co., US Patent 4 415 690, 15 November 1983.
44. GRIMM, D. C., Goodyear Tire & Rubber Co., US Patent 4 452 936, 5 June 1984.
45. SHAHEEN, F. G. and GRIMM, D. C., 'Carboxylated nitrile rubber', *ACS Meeting, Los Angeles*, 24 April 1985.
46. *Chemigum NX 775*, Technical Leaflet, Goodyear Chemical Division, Akron, Ohio, 1985.
47. SHELL, R. L., 'Carboxylated nitrile rubber and its use in oil field applications', Southern Rubber Group, Fort Worth, Texas, 18 February 1982.
48. WALTER, B. W. ed., *Handbook of Thermoplastic Elastomers*, Van Nostrand–Reinhold Co., New York, 1979.
49. GESSLER, A. M., US Patent 3 037 954, 1962.
50. CORAN, A. Y. and PATEL, R., *Rubber Chem. Technol.*, **53**, 1980, 141.
51. CORAN, A. Y. and PATEL, R., *Rubber Chem. Technol.*, **54**, 1981, 892.
52. ABDOU-SABET, S., WANG, Y. L. and CHU, E. F., 'Geolast—a new oil resistant thermoplastic rubber', paper presented to *128th ACS Rubber Div. Meeting, Cleveland, OH*, 3 Oct. 1985.
53. GAYLORD, N. G., *Adv. Chem. Ser.*, **142**, 1975, 76.
54. HOFFMAN, W., *Ind. Plast. Mod.*, 1961, 37.
55. PEDLEY, K. A., *Polym. Age*, May 1970, 47.

56. SHARP, T. J. and ROSS, J. A., *Trans. Inst. Rubber Ind.*, **37**, 1961, 157.
57. HORVATH, J. W., WILSON, W. A., LINDSTROM, H. S. and PURDON, J. R., *Appl. Polym. Symp.*, **7**, 1968, 95.
58. SCHWARZ, H. F., *Elastomerics*, **112**(11), 1980, 17.
59. SCHWARZ, H. F., *Rubber World*, **187**(2), 1982, 22.
60. GIUDICI, P. and MILNER, P. W., 'PVC/NBR blends', paper presented to *PRI Conference, Brussels*, 1976.
61. 'Einsatz von NBR zur PVC-Modifizierung', *Gummi Fasern Kunstoffe*, **381**(2), 1985, 62.
62. *The Synthetic Rubber Manual*, 9th edn, International Institute of Synthetic Rubber Producers Inc., Houston, Texas, USA, 1983.
63. BLEY, J. W. F. and SCHWARZ, H. F., 'Design of alloys of PVC with NBR copolymers to produce thermoplastic elastomers', Paper No. 93, *128th ACS Rubber Div. Meeting, Cleveland, Ohio*, October 1985.
64. *Chemigum P83*, Technical Leaflet, Goodyear Chemical Division, BP 31, Les Ulis, France, 1984.
65. '*Elastar*'—*PVC/NBR Thermoplastic Polymers*, Nippon Zeon Technical Literature, 1985.
66. *Hycar VT Elastomers*, Technical Literature, B. F. Goodrich Chemical Division, Cleveland, Ohio, 1985.
67. MILNER, P. W., *Gummi Asbest. Kunst.*, **25**(7), 1972, 634.
68. MILNER, P. W., 'Obtaining the maximum physical properties from powdered rubber in a one-step process', paper presented to *4th Europ. Conf. Plastics and Rubbers, Paris*, 1974.
69. LEHMEN, J. P., *Europ. Rubber J.*, **164**(1), 1982, 53.
70. LEHMEN, J. P., 'Powdered rubber technology in technical goods production', *PRI Symposium, Southampton University*, April 1978.
71. LEHMEN, J. P., *Gummi. Asbest. Kunst.*, **35**(5), 1982, 253.
72. DRAKE, R. S. and MCCARTY, W. J., *Rubber World*, **159**(1), 1968, 51.
73. POCIUS, A. V., 'Elastomer modification of structural adhesives', *Rubber Chem. Technol.*, **58**, 1985, 622.
74. ELDRED, R. J., *Rubber Chem. Technol.*, **57**, 1984, 320.

Chapter 3

EPOXIDIZED NATURAL RUBBER

C. S. L. Baker and I. R. Gelling

*Malaysian Rubber Producers' Research Association, Tun Abdul Razak
Laboratory, Brickendonbury, Hertford, UK*

1. INTRODUCTION

The mechanical properties of natural rubber (NR) are generally superior to those of synthetic rubbers. However, NR cannot compete with the speciality synthetic elastomers with regard to such properties as gas permeability and oil resistance.

It has been recognized for many years that chemical modification of NR can be utilized to produce new materials. Chlorinated,[1] hydrochlorinated[2] and cyclized[3] NR have all been produced commercially and poly(methyl methacrylate) grafted[4] and depolymerized NR[5] are still available. More recent chemical reactions have been restricted to limited levels of modification to change selected properties while retaining the overall rubbery nature of the polymer.[6] These modifications have also been utilized for antioxidant functions,[7] vulcanization sites[8] and coupling to silica fillers.[9]

The epoxidation of NR and other unsaturated polymers has been reported in the literature.[10] However, there are little data on the properties of these materials and, where they do exist, they tend to be conflicting. The above reactions established criteria for the chemical modification of NR, which led to the development of clean epoxidized natural rubber (ENR). These new polymers have improved oil resistance and decreased gas permeability, whilst retaining many of the properties of NR and also exhibiting some novel features.

The object of this chapter is to describe the properties of ENR and to suggest applicational areas.

2. POLYMER MANUFACTURE

2.1. Epoxidation Chemistry

The chemistry of epoxidation of unsaturated compounds is well documented in the literature[11] and is normally carried out by employing acetic or formic peroxy acids, either preformed or generated *in situ*. The *in-situ* peroxy formic acid route is illustrated in Fig. 1. The mole ratio of formic acid to hydrogen peroxide is normally 0·2–0·5 as the former reagent is recycled within the reaction sequence.

If the weaker acetic acid is used, then an additional acid catalyst, either sulphuric acid or a sulphonic acid ion exchange resin, is necessary to promote the peroxy acid formation (reaction (1)). The epoxidation (reaction (2)) is generally rapid, but conditions need to be carefully controlled to avoid secondary epoxide ring-opening reactions. Two types of ring-opening reaction can occur (Fig. 2). Simple isolated epoxide groups[12] yield glycols, hydroxyesters and derived products (reaction (3)), but in polyunsaturated compounds, where there are adjacent epoxide groups, the major product is a cyclic ether (reaction (4)).[13,14]

2.2. Production

Epoxidized natural rubber (ENR) is produced in Malaysia from NR latex employing the hydrogen peroxide/formic acid *in-situ* route (Fig. 3) with the reaction conditions carefully controlled to avoid secondary

$$H-C\overset{O}{\underset{OH}{\diagup}} + H_2O_2 \longrightarrow H-C\overset{O}{\underset{O-OH}{\diagup}} + H_2O \qquad (1)$$

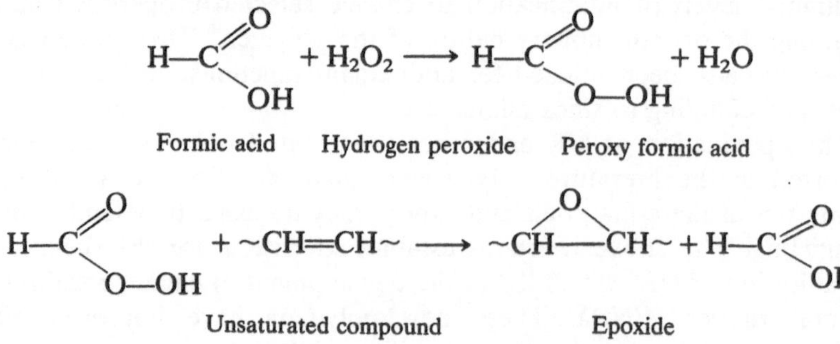

Formic acid   Hydrogen peroxide   Peroxy formic acid

$$H-C\overset{O}{\underset{O-OH}{\diagup}} + \sim CH=CH\sim \longrightarrow \sim \overset{O}{CH-CH}\sim + H-C\overset{O}{\underset{OH}{\diagup}}$$

Unsaturated compound   Epoxide   (2)

FIG. 1. *In situ* epoxidation with formic acid and hydrogen peroxide.

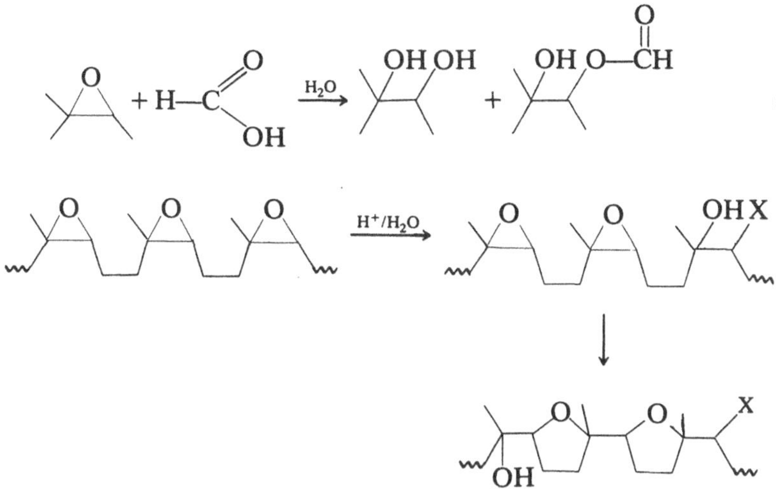

$$(3)$$

$$(4)$$

FIG. 2. Epoxide ring-opening reactions.

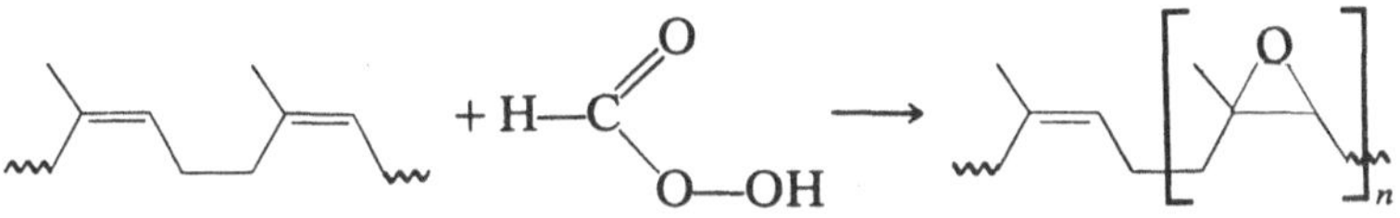

| NR field latex | Peroxy formic acid (formed *in situ*) | ENR-25 ($n = 0.5$) ENR-50 ($n = 1$) |

FIG. 3. Production of epoxidized natural rubber.

ring-opening reactions. The NR latex is first stabilized against acid coagulation by the addition of a non-ionic surfactant and the epoxidation carried out at 60–70°C.[15] The epoxidized latex is neutralized prior to heat coagulation and thoroughly washed before drying in hot air. Two materials, 25 mol% epoxidized natural rubber (ENR-25) and 50 mol% epoxidized natural rubber (ENR-50), are currently produced.

3. POLYMER STRUCTURE

3.1. Epoxide Analysis
The epoxide content of ENR-25 and ENR-50 is within 1–2% of the quoted values as determined by a combination of elemental oxygen

analysis and proton nuclear magnetic resonance (^{1}H NMR) spectroscopy. The latter technique also clearly shows that the modification is
essentially 'clean', i.e. only epoxide groups have been introduced into
the NR molecule.[16] Any epoxide ring-opened structures are less than
1·0 mol%, the limits of detection of the analytical techniques
employed.

3.2. Stereochemistry
NR has a stereoregular *cis*-1,4-polyisoprene structure. The epoxidation reaction with peroxy acids is stereospecific[17] and thus ENR retains
the stereoregular all-*cis*-1,4-configuration of NR.

3.3. Epoxide Distribution
The distribution of epoxide groups along the polymer backbone will to
a large extent determine the physical properties of the ENR concerned. If it is assumed that the epoxidation of unsaturated NR units is
random, and not controlled by chemical or steric factors, then the
mole percentages of the variously positioned epoxide groups for any
total level of modification may be calculated (Fig. 4). The epoxide
group sequences of both ENR-25 and ENR-50 have been measured by

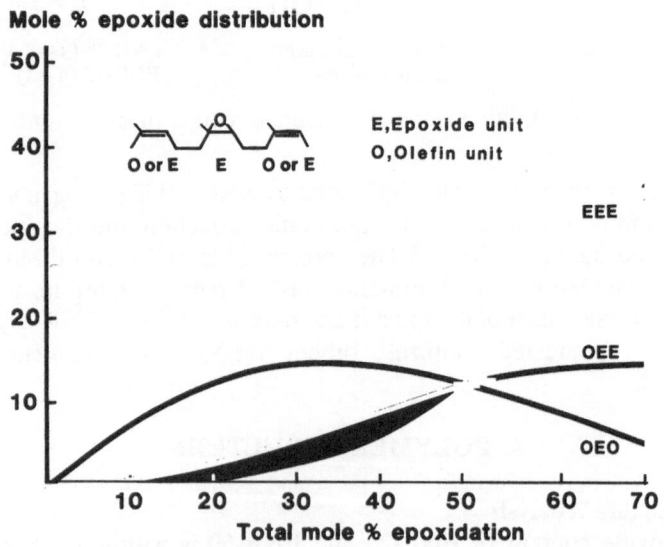

FIG. 4. Calculated epoxide sequences for randomly epoxidized NR.

^{13}C NMR spectroscopy and the observed sequences agree with those calculated for a totally random epoxidation.[16]

Chemical titrimetric analysis of epoxide groups in ENR, e.g. the method of Jay[18] or Swern,[19] gives lower epoxide contents than spectroscopic techniques[20] and is not a reliable measure of the total epoxide content of these polymers. Where there are two or more adjacent epoxide groups these reagents initiate the ring-opening of the epoxide sequence (reaction (4), Fig. 2), and thus are more a measure of the epoxide distribution than the total epoxide content.[20] Epoxidized NRs are thus randomly epoxidized *cis*-1,4-polyisoprenes.

4. RAW RUBBER PROPERTIES

All grades of ENR are prepared from premium-quality NR latex and are light yellowish-brown in colour. They are supplied as standard 33·3 kg polythene-wrapped bales.

4.1. Mooney Viscosity and Solubility

Both ENR-25 and ENR-50 have Mooney viscosities in the range 75–90 on production and for certain applications may need to be premasticated prior to compounding.

Epoxidation of NR increases its polarity, and thus the solubility of ENR depends on the level of epoxidation and the nature of the solvent. In order to ease solution ENRs should be premasticated. Recommended solvents are toluene, chloroform and tetrahydrofuran.

4.2. Storage

At low temperatures NR crystallizes and once this has occurred the rubber has to be thawed before it can be mixed. Epoxidation, even at low levels (Fig. 5), reduces the rate of low-temperature crystallization.[21] At epoxide levels of 25 and 50 mol% the rubbers are relatively resistant to low-temperature crystallization.

4.3. Glass Transition Temperature

For every 1 mol% epoxidation, the glass transition temperature (T_g) increases by approximately 1°C (Fig. 6); thus ENR-25 has a T_g of −47°C and ENR-50 a T_g of −22°C. The effect of this change in T_g is clearly seen in the properties of these materials which are described in subsequent sections.

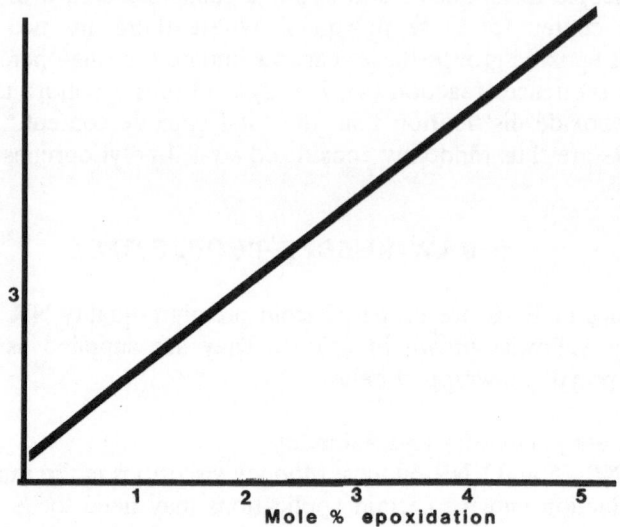

FIG. 5. The effect of epoxidation on the low-temperature crystallization of NR.

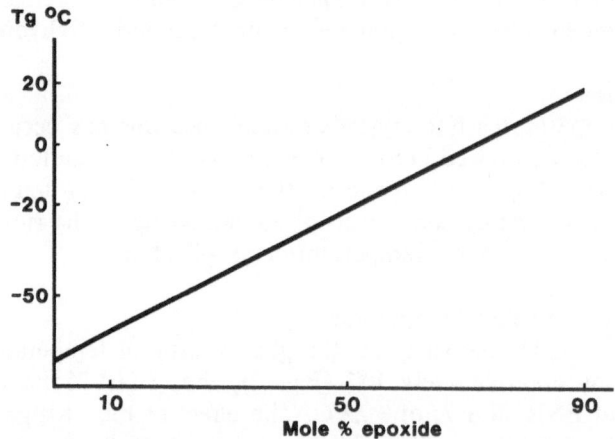

FIG. 6. Change of glass transition temperature with epoxide content of ENRs.

These transitions are all sharp events and are thus an indication of the homogeneity of the rubbers.

5. COMPOUNDING AND VULCANIZATION

5.1. Compounding

The Mooney viscosity and the need to ensure that ENR is kept basic are the main factors to be considered during compounding. On production the viscosities are typically 75–90 and ENR is slightly basic.

5.1.1. Mastication

Prior to compounding, particularly if to be blended with lower viscosity polymers, ENR may need to be premasticated.

Mastication of ENR can be carried out on an open mill or in an internal mixer, and chemical peptizers may be employed to increase the rate of breakdown (Fig. 7).

5.1.2. Black Mixing

The incorporation of blacks into ENR can result in large increases in Mooney viscosity.[22] With the coarser low structure blacks, any increase in viscosity is no greater than that observed with many other polymers. However, high loadings of the more reinforcing blacks, such as HAF(N330), ISAF(N220) or SAF(N110), can result in high rises in viscosity during black mixing, which is associated with a second peak in the mixing curve (Fig. 8). The addition of a base substantially inhibits this phenomenon (Fig. 8), and, as produced, ENR should be sufficiently basic. However, the air ageing properties of ENR depend on its basicity and it may be necessary to add additional base to optimize this property. This will be dealt with in detail in a later section. If a base, e.g. sodium carbonate, calcium stearate, magnesium oxide, calcium oxide or sodium acetate, is to be included in the formulation then it should be added early in the mix cycle.

5.1.3. Mix Cycle

The following schedule should be followed in order to obtain satisfactory mixing characteristics.

Stage 1 Mastication of ENR with addition of powders and base.

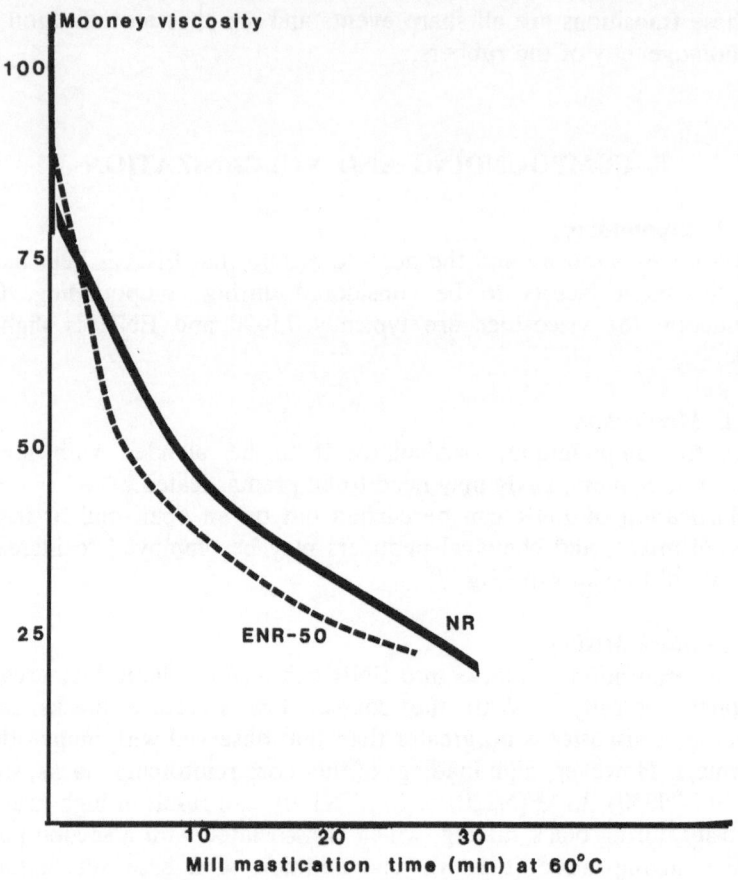

FIG. 7. Mill breakdown behaviour of ENR-50 compared with NR at 60°C.

Stage 2 Black incorporation. The exact cycle will depend on the black loading and oil, but equivalent cycles to those employed for NR are normally satisfactory.

Stage 3 Addition of curatives, either in the internal mixer or on an open mill.

5.2. Vulcanization
ENR can be vulcanized by any of the standard sulphur or peroxide formulations normally employed for unsaturated polymers.[23]

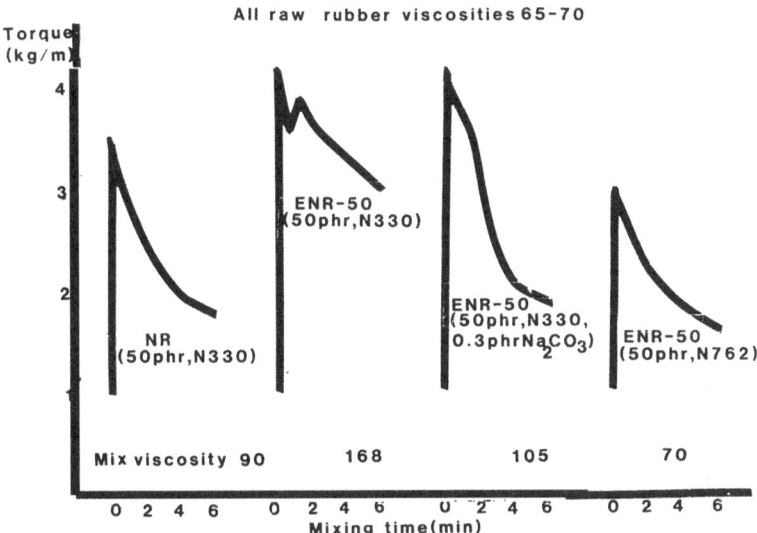

FIG. 8. Mixing characteristics of ENR-50 and blacks (+5 phr aromatic oil) in the Brabender Plasti-Cord, 100 rpm, initial temperature of 100°C.

5.2.1. Sulphur Formulations

Although acceptable cure characteristics are obtained with a conventional system based on 2·5 phr sulphur and 0·5 phr sulphenamide, this formulation is not recommended for ENR as the vulcanizates have poor ageing characteristics compared with unmodified NR. The cause of this is discussed in Section 6.9. Semi-EV or EV-type cure systems have been found to be the most satisfactory for ENR. Examples of such systems are recorded in Table 1. The need to compound ENR with a base affects the vulcanization characteristics, and one of the factors to be considered in the choice of base is the processing safety of the mix.[24] Stronger bases can markedly reduce scorch delay by base catalysis of the vulcanizing system and should only be employed at low levels, whereas the less basic compounds can be employed at higher concentrations (Fig. 9).

The Mooney scorch times of ENR mixes can be increased by the use of commercial prevulcanization inhibitors (PVI) once the base has been added. The response of these materials depends on the cure system employed and parallels that observed with NR and other unsaturated polymers.

TABLE 1
COMPOUNDING FORMULATIONS FOR ENR (phr)

	Semi-EV	*EV*	
ENR	100	100	100
Base	0·3–5		
Filler		as required	
Aromatic oil	5	5	5
Zinc oxide	5	5	5
Stearic acid	2	2	2
Antioxidant[a]	2	2	2
Sulphur	1·5	0·3	0·8
MBS[b]	1·5	2·4	3·0
TMTD[c]	—	1·6	—
CTP[d]	0·2	0·2	0·2

[a] Poly-2,2,4-trimethyl-1,2-dihydroquinoline or *p*-phenylenediamine type.
[b] *N*-Oxydiethylene benzothiazole-2-sulphenamide.
[c] Tetramethylthiuram disulphide.
[d] *N*-(Cyclohexylthio)phthalimide or alternative pre-vulcanization inhibitor.

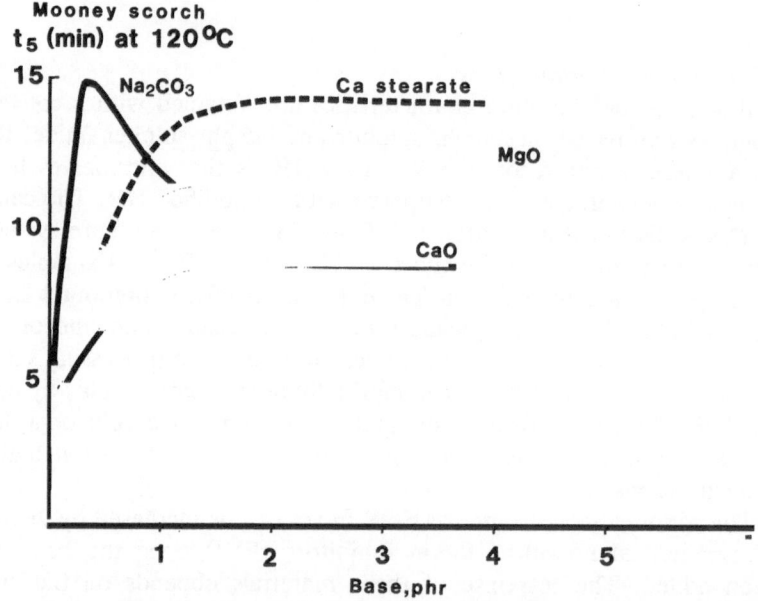

FIG. 9. The effect of bases on the Mooney scorch properties of a semi-EV (1·5 S; 1·5 MBS; 30 phr N330 black) compound.

The vulcanization profiles of a semi-EV black filled (30 phr ISAF, N220) ENR-50 compound (Table 1) with and without the addition of a prevulcanization inhibitor are recorded in Fig. 10. The overall response of ENR to changes in accelerator type and level is similar to that observed with NR. The major adjustment that has to be considered is due to the presence of base.

5.2.2. Mechanism of Sulphur Vulcanization
The vulcanization by sulphur alone is faster for ENR than for NR as isolated double bonds react more rapidly than contiguous double bonds. The presence of isolated double bonds also increases the efficiency of vulcanization as they inhibit the formation of in-chain cyclic sulphides.[25]

In the presence of sulphenamide accelerators little difference is observed between ENR and NR.[25] Similar types of crosslink, mono-, di- and poly-sulphidic, are produced and their ratio varies with the sulphur/accelerator ratio as with other highly unsaturated polymers.

5.2.3. Peroxide Vulcanization
Peroxide cure systems can be utilized to crosslink ENR and some examples of formulations that have been investigated, and their

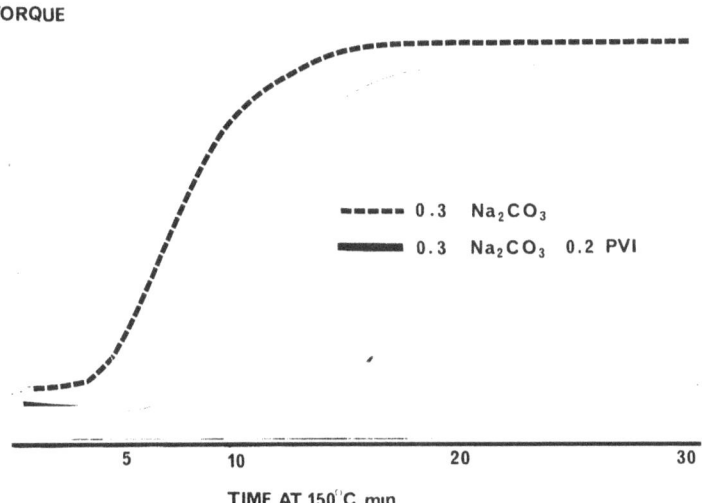

FIG. 10. Vulcanization characteristics of ENR-50.

TABLE 2
ENR PEROXIDE VULCANIZATES

Formulation (phr)			
ENR-50	100	100	100
SRF black (N762)	30	30	30
DOS (dioctylsebacate)	10	10	10
Flectol H	2	2	2
Dicumyl peroxide	3·0	1·0	1·0
HVA2 (N,N'-metaphenylenedimaleimide)	—	3·0	3·0
NDPA (4-nitrosodiphenylamine)	—	—	0·2
Cure characteristics at 160°C (Monsanto Rheometer: arc ±1)			
t_{s1} (min)	2·5	1·5	3·4
t_{95} (min)	29·6	20·3	21·6
ΔT (in lb)	13·9	18·2	17·3
Properties			
Unaged			
Modulus at 100% (MPa)	0·86	1·56	1·21
Modulus at 300% (MPa)	6·4	—	—
Tensile strength (MPa)	10·1	10·7	10·8
Elongation at break (%)	364	262	273
Compression set (24 h/70°C, 25%)(%)	15	14	16
Air-aged (5 days/100°C)			
Modulus at 100% (MPa)	0·86	1·74	2·02
Tensile strength (MPa)	10·3	9·2	9·4
Elongation at break (%)	368	238	236

characteristics, are recorded in Table 2. It is assumed that crosslinking occurs via the residual double bonds by a similar mechanism to that with NR. Tensile properties are generally inferior to corresponding sulphur vulcanizates and the improvement in compression set obtained with NR is not so apparent with ENR. However, as with NR, ageing properties are superior.

6. PHYSICAL PROPERTIES

6.1. Strain Crystallization
An X-ray diffraction study[26] of gum ENR vulcanizates shows that these materials undergo strain crystallization. The crystallinity is clearly seen in Fig. 11 and the degree of crystallinity (Table 3) was

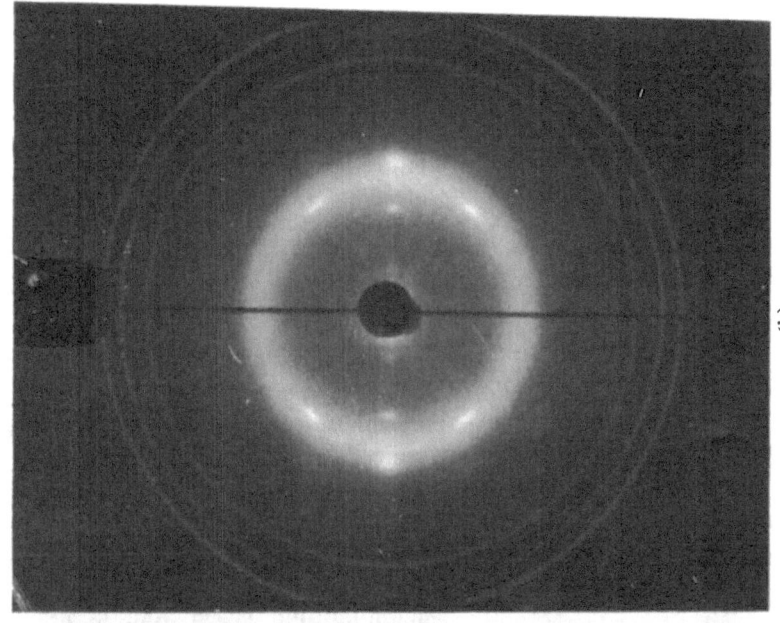

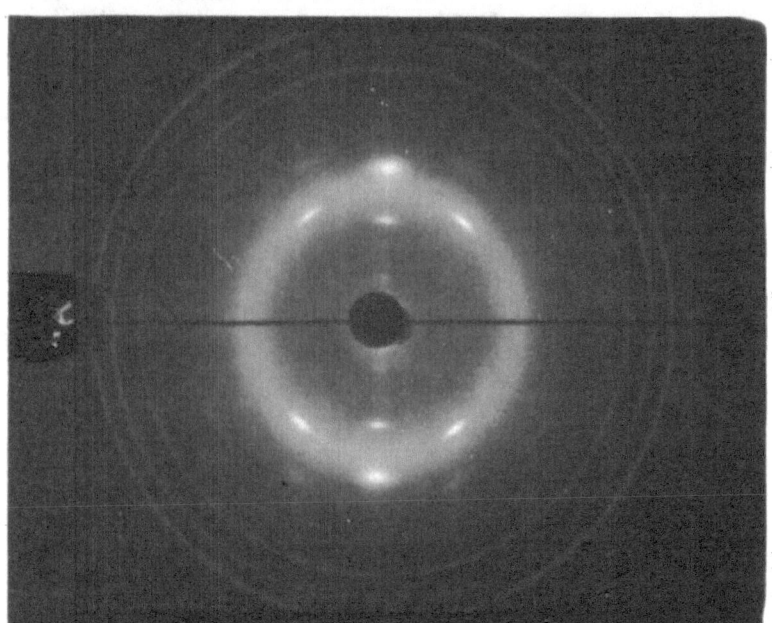

(a)

(b)

Fig. 11. X-ray diffraction patterns from (a) NR and (b) ENR-50.

TABLE 3

DEGREE OF CRYSTALLINITY OF NR, ENR-25 AND ENR-50

Rubber	Crystallinity (%)[a]
NR	11
ENR-25	11
ENR-50	10

[a] Samples extended to 400% at 20°C.

determined from the variation of peak intensities of the amorphous halo and is equal to that of NR under these experimental conditions.

The unit cell of *cis*-1,4-polyisoprene has been described as orthorhombic or monoclinic. At first sight it might be surprising that the epoxide group can be incorporated in the crystal lattice of the strained polymer. A *trans*-isoprene unit is incorporated only with great difficulty, if at all. However, the *cis*-configuration is rigorously maintained by the epoxidation reaction and the oxygen atom occupies a relatively inconspicuous volume. The unit cell volume of ENR-50 has increased to $0.999 \, nm^3$ from the original NR value of $0.955 \, nm^3$.

Beyond 50 mol% epoxidation a rapid decrease in strain crystallization was observed.

6.2. Strength Properties

6.2.1. Gum Vulcanizates

The ability of ENR to undergo strain crystallization is reflected in its high tensile strengths and tear properties which are comparable with those of NR (Table 4).

The tear strengths of ENR are illustrated in Fig. 12, where it can be seen that, over a range of tearing rates, ENR exhibits high values in contrast to a non-crystallizing rubber such as a styrene–butadiene copolymer (SBR).[27]

6.2.2. Black Filled Vulcanizates

The properties of black filled ENR vulcanizates are recorded in Table 5 and, as expected, tensile strengths are high. However, reinforcement with black fillers does not increase the tear strength of ENR

TABLE 4

COMPARISON OF NR, ENR-25 AND ENR-50 GUM VULCANIZATE PROPERTIES

Formulation (phr)			
Polymer	100		
Zinc oxide	5		
Stearic acid	2		
Antioxidant[a]	2		
Sulphur	1·5		
MBS[b]	1·5		

Property	NR(SMRL)	ENR-25	ENR-50
Modulus at 100% (MPa)	0·68	0·69	0·74
Modulus at 300% (MPa)	1·46	1·50	1·56
Tensile strength (MPa)	28·7	24·3	28·3
Elongation at break (%)	760	770	770
Hardness (IRHD)	35	34	36
Tension fatigue, ring			
0–100% extn (kcs)[c]	145	134	234
50–150% extn (kcs)[c]	1300	915	880
Tear strength, trouser (N/mm)	6·8	6·9	6·3

[a] Poly-2,2,4-trimethyl-1,2-dihydroquinoline or p-phenylenediamine type.
[b] N-Oxydiethylenebenzothiazole-2-sulphenamide.
[c] kcs = kilocycles to failure.

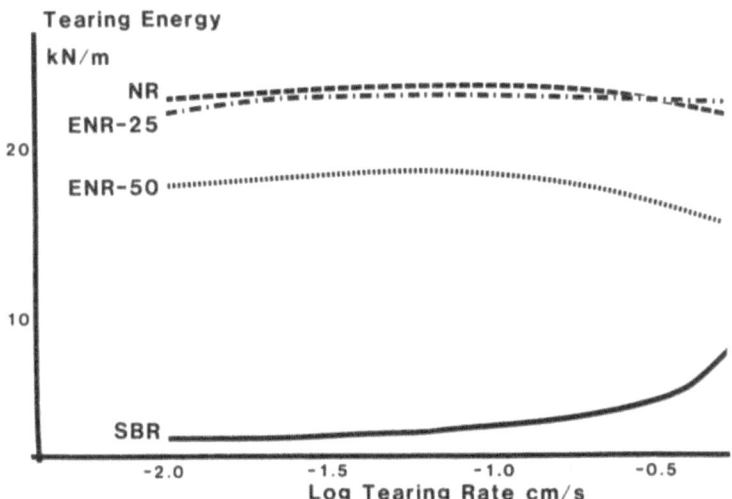

FIG. 12. Comparative tearing energies of NR, ENR-25, ENR-50 and SBR vulcanizates.

TABLE 5
PHYSICAL PROPERTIES OF BLACK FILLED ENR VULCANIZATES

Formulation (phr)						
NR (SMRL)	100	100	—	—	—	—
ENR-25	—	—	100	100	—	—
ENR-50	—	—	—	—	100	100
Sodium carbonate	—	—	0·3	0·3	0·3	0·3
ISAF (N220)	30	30	30	30	30	30
Aromatic oil	5	5	5	5	5	5
Zinc oxide	5	5	5	5	5	5
Stearic acid	2	2	2	2	2	2
Antioxidant	2	2	2	2	2	2
Sulphur	1·5	0·3	1·5	0·3	1·5	0·3
MBS	1·5	2·4	1·5	2·4	1·5	2·4
TMTD	—	1·6	—	1·6	—	1·6
PVI	—	—	0·2	—	0·2	—
Properties						
Mooney scorch, 120°C (min)	48	23	30	18	25	14
Time opt. cure, 150°C (min)	24	45	17	40	23	22
Tensile strength (MPa)	32·6	27·0	28·4	25·6	28·3	27·3
Elongation at break (%)	600	550	590	550	585	565
Modulus at 100% (MPa)	1·53	1·47	1·83	1·37	1·95	1·83
Modulus at 300% (MPa)	7·7	7·7	8·9	6·7	8·8	8·7
Hardness (IRHD)	56	59	57	52	64	58
Resilience, Dunlop 23°C (%)	76	78	66	59	19	24
Compression set (24 h/70°C, 25%) (%)	21	17	22	15	30	17
Tension fatigue, ring						
0–100 (kcs)[a]	103	65	165	60	250	97
50–150% (kcs)[a]	>1560	1300	1200	615	550	317
Goodrich HBU 0·225 in stroke,						
24 lb load from 23°C (ΔT°C)	40	43	40	45	42	51
Abrasion, Akron (mm³/500 rev)	22	44	14	15	12	12

[a] kcs = kilocycles to failure.

vulcanizates in the same manner as is observed with NR.[27] Higher black loadings are necessary to achieve comparable tear strengths (Fig. 13). A marked reduction in non-relaxing fatigue properties is also observed as a result of the modification. However, it should be noted that an ENR-50 vulcanizate filled with 30 phr ISAF(N220) black has a non-relaxing fatigue life nearly ten times longer than that of a non-crystallizing rubber such as NBR.

6.3. Oil Resistance

The polarity of ENR increases with rise in epoxide content and hence the pattern of oil/solvent resistance changes. At high epoxidation levels, these materials become more resistant to hydrocarbons but their resistance to polar solvents decreases (Table 6).

TEAR

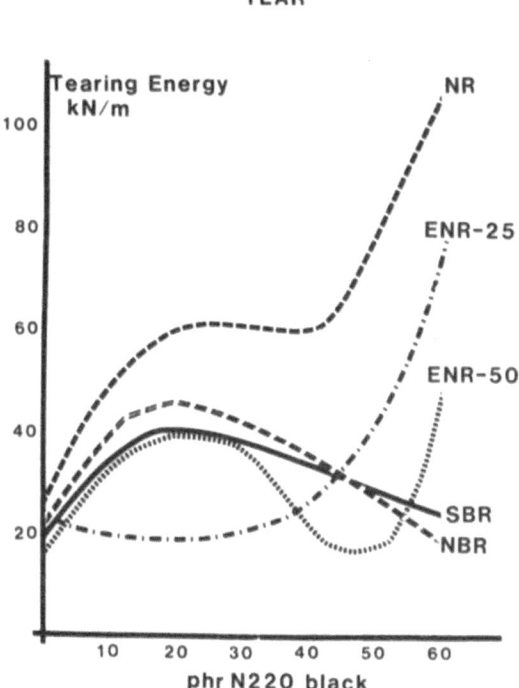

Fig. 13. The effect of black (ISAF N220) loading on the tear properties of NR, ENR-25, ENR-50, SBR and NBR.

TABLE 6
PERCENTAGE VOLUME SWELLING OF POLYMERS IN VARIOUS FLUIDS

	NR	ENR-25	ENR-50	NBR	CR
70 h/100°C in					
ASTM No. 1 oil	97	11	−5	−4	2·5
ASTM No. 2 oil	145	71	13	7	33
ASTM No. 3 oil	209	126	36	18	59
70 h/23°C in					
ASTM ref. fuel B	178	127	59	38	—
Ethanol	−0·45	15	28	16	—
Methyl ethyl ketone	−36	51	98	178	—

The resistance of ENR-50 to ASTM oils approaches that of a medium (34% acrylonitrile) NBR copolymer and is superior to that of polychloroprene rubber (CR). Hence, ENR-50 can be considered as a replacement for these polymers where oil resistance is necessary. ENR-50 does not show any advantage over NBR with regard to resistance to peroxidized or alcohol-containing fuels.

6.4. Air Permeability
Epoxidation also results in a substantial decrease in air permeability. Figure 14 records the permeation constants of various grades of ENR over a range of temperatures in comparison with those of NR and butyl rubber. The air permeability of ENR-50 is similar to that of butyl rubber and thus it can be used in applications requiring low air permeability.

6.5. Hysteresis
Typical resiliences at ambient temperatures of ENR-25 and ENR-50 are recorded in Table 5. These values are a consequence of the rise in glass transition temperature on epoxidation and are thus highly temperature-dependent. Figure 15 records the temperature response of ENR resilience. Blends of ENR (Fig. 15) modify the temperature response and more practical systems for damping applications are given in Section 7.2.

6.6. Silica Filled ENR
To optimize the reinforcement properties of silica in NR and other unsaturated polymers, a silane coupling agent has to be employed.

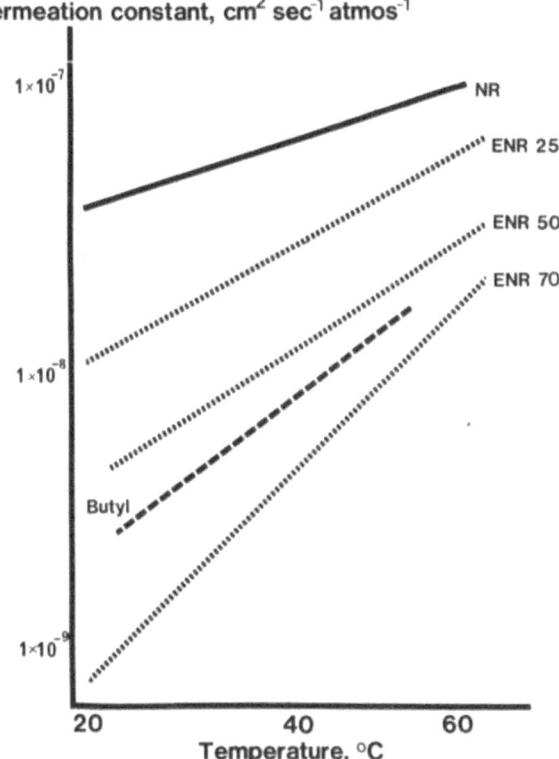

FIG. 14. Air permeability of various grades of ENR.

With ENR-25 and ENR-50 high degrees of reinforcement are achieved with silica fillers in the absence of coupling agents. Table 7 records the properties of ENR vulcanizates filled with 50 phr of silica (no coupling agent) and a corresponding loading of a N330 carbon black. It can be clearly seen that comparable reinforcement is obtained in the case of ENR-25 and ENR-50 whereas, with silica filled NR, a much lower modulus and higher compression set, heat build-up and abrasion values are obtained compared with the black filled vulcanizate.

6.7. Bonding
The increase in polarity of ENR results in changes in the adhesive characteristics.

% Rebound

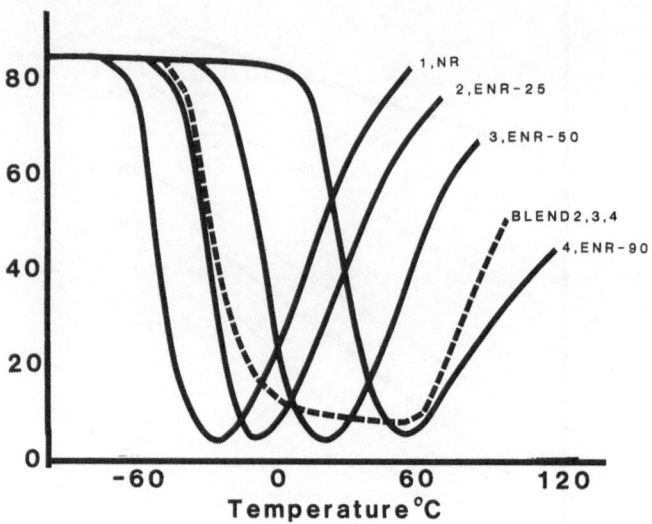

FIG. 15. Rebound resilience of ENR.

6.7.1. Tack

With increasing levels of epoxidation, the tack of ENR decreases towards non-polar polymers such as NR or SBR whilst ENR becomes more compatible with polar polymers. The tack of ENR-50 and blends with NR is recorded in Table 8.

6.7.2. Cured Adhesion

Values of cured adhesion of ENR-25 and ENR-50 measured by a 180° peel test are recorded in Table 9. These results also reflect the change in polarity on epoxidation; adhesion to non-polar polymers such as NR is low, but as the polarity of the rubber increases high adhesion is observed.

A number of rubber products consist of rubber/PVC composites (e.g. rubber-covered PVC conveyor belting) and the high adhesive strength of ENR-50 to PVC confers obvious benefits to articles with this type of construction since failure often occurs at the interface. An ENR-50 compound could also be used to bond other polymers to PVC.

TABLE 7
COMPARISON OF BLACK AND SILICA FILLED VULCANIZATES

	NR		ENR-25		ENR-50	
	Black	Silica	Black	Silica	Black	Silica
Formulation (phr)						
Rubber	100	100	100	100	100	100
Sodium carbonate	—	—	0·3	0·3	0·3	0·3
Black (N330)	50	—	50	—	50	—
Silica (Hi-Sil 233)	—	50	—	50	—	50
Process oil	4	4	4	4	4	4
Zinc oxide	5	5	5	5	5	5
Stearic acid	2	2	2	2	2	2
Antioxidant	2	2	2	2	2	2
Sulphur	2	2	2	2	2	2
MBS	1·5	1·5	1·5	1·5	1·5	1·5
DPG (diphenylguanidine)	—	0·5	—	0·5	—	0·5
Properties						
Hardness (IRHD)	65	69	69	67	73	68
Modulus at 300% (MPa)	11·9	5·8	12·4	12·8	13·5	12·6
Tensile strength (MPa)	29·4	23·2	25·4	21·0	24·5	22·4
Elongation at break (%)	495	720	435	405	500	435
Abrasion, Akron (mm^3/500 revs)	21	63	14	15	11	14
Abrasion, DIN (mm^3)	199	364	272	250	278	289
Compression set (24 h/70°C, 25%)(%)	18	32	17	18	21	22
Tension fatigue, ring 0–100% (kcs)	70	51	65	52	93	58
Goodrich HBU, 30 min at 100° (ΔT°C)	7	47	7	7	23	19

TABLE 8
TACK PROPERTIES OF ENR AND ENR/NR BLENDS

Polymer	Tack to NR (N/mm)	Cohesive tack (N/mm)
ENR-50	0·66	9·8
ENR-50/NR (80/20)	1·95	—
ENR-50/NR (70/30)	2·7	6·1

6.7.3. Steel Bonding

High bond strengths are obtained between ENR and steel using a standard Chemlok (205, 220) two-coat system with the steel sandblasted and degreased prior to bonding. Samples tested at a peel angle of 90° had bond strengths in excess of 9·5 N/mm for both ENR-25 and ENR-50 in a soluble EV formulation (S, 0·5; MBS, 1·2; TBTD, 0·5 phr). In both cases failure occurred in the rubber, and not at the bond.

TABLE 9
CURED ADHESION OF ENR TO OTHER POLYMERS

Ply composition	Adhesive strength (N/mm)
ENR-50/NR	0·8[a]
ENR-25/NR	6·6[a]
ENR-50:NR (60:40)/NR	9·8[a]
ENR-50:NR (60:40)/NR; SBR carcass stock	10·4[a]
ENR-25/NBR	>10[b]
ENR-50/NBR	>10[b]
ENR-25/CR	>12[b]
ENR-50/CR	>12[b]
NR/PVC	0·4[a]
ENR-25/PVC	2·4[a]
ENR-50/PVC	7·8[b]
NR/PVC[c]	6·5[b]

[a] Bond failure.
[b] Rubber failure.
[c] A blend of NR/ENR-50(50:50) was used as an intermediate layer or applied as a solvent-based adhesive coat.

TABLE 10
WET GRIP OF RUBBERS ON A SMOOTH
CONCRETE SURFACE

Rubber	Wet grip[a]
NR	64
SBR	73
ENR-25	90
ENR-50	100
OESBR	79
OENR	81

[a] The ENR-50 sample was taken as 100 and the other results rated accordingly.

6.8. Wet Grip

The wet grip properties of rubbers play a vital role in such applications as tyres, shoe soles and floor coverings. ENR exhibits high wet grip characteristics as measured by the laboratory pendulum test (Table 10); they are in excess of those of NR and SBR and even of their oil-extended versions.

6.9. Air Ageing

ENR possesses fewer double bonds than NR and therefore would be expected to be more resistant to oxidation. However, the oven air ageing of a conventionally cured ENR vulcanizate (S, 2·5; sulphenamide accelerator, 0·5 phr) was found to be poor. A rapid hardening occurred on air ageing. This oxidative hardening is not an intrinsic property of ENR since when peroxide and efficient vulcanizing systems (S, 0·3; sulphenamide, 2·4; TMTD, 1·6 phr) were used their ageing characteristics were observed to be similar to those of the corresponding NR vulcanizates (Fig. 16).

The anaerobic ageing of both conventional and efficient vulcanizates of ENR is comparable with that of the corresponding NR vulcanizate. The rapid hardening of conventional sulphur vulcanizates was reflected in an increased modulus, loss in tensile strength and reduction in elongation at break. Although the air ageing of efficient vulcanizates of ENR and NR are more similar, the former still ages via a hardening mechanism. Analysis of aged conventional ENR-50 vulcanizates has

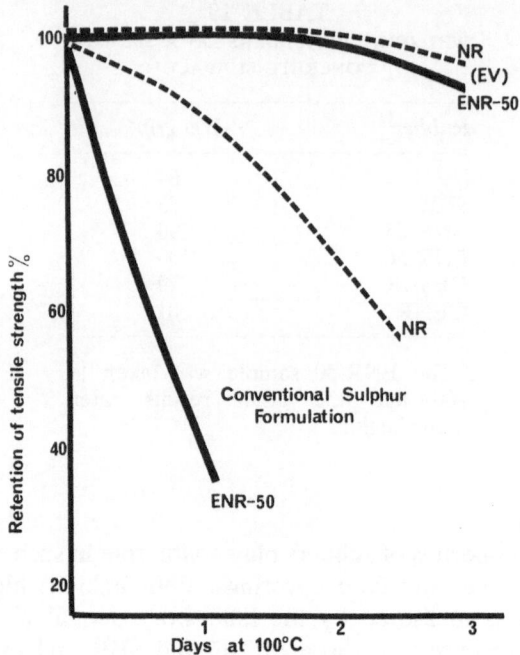

FIG. 16. Ageing of ENR vulcanizates.

recorded a reduction in the number of epoxide groups, and an increase in glass transition temperature and crosslink density.

Model studies[25] have shown that the ageing of ENR sulphur vulcanizates occurs by acid-catalysed ring-opening of epoxide groups with the formation of ether crosslinks. The acids are produced by the thermal decomposition of oxidized sulphides.

$$\text{Sulphide crosslink} \xrightarrow[\text{Heat}]{O_2} \text{sulphur-containing acid}$$

$$\downarrow \text{epoxide group}$$

$$\text{Ring-opening, ether crosslink}$$

The addition of a base neutralizes the acids produced during ageing and substantially improves the ageing properties of ENR (Table 11). Sodium carbonate is the most effective base but significant amounts can result in 'scorch' problems (Fig. 9). To control the scorch a prevulcanization inhibitor can be used in conjunction with sodium

TABLE 11
THE AIR AGEING OF ENR-50 VULCANIZATES^a IN THE PRESENCE OF BASE

Property	Retention after air ageing 3 days/100°C (%)		
Base:	None	Sodium carbonate (2 phr)	Calcium stearate (5 phr)
Modulus at 100%	180	104	119
Tensile strength	78	96	86
Elongation at break	60	100	89

[a] EV formulation (S, 0·3; MBS, 2·4; TMTD, 1·6 phr).

carbonate. Alternatively the salt of an organic acid (e.g. calcium stearate) may be employed since, although these materials are not quite as effective as sodium carbonate, they do not affect scorch to the same extent.

6.10. Ozone Resistance

The ozone resistance of ENR-25 is comparable with that of NR and it shows a similar response to waxes and chemical antiozonants. Uncompounded ENR-50 has superior ozone resistance, but it does not respond as well to antiozonants, and the ozone resistance of protected ENR-50 vulcanizates is inferior to that of NR and ENR-25 at the same level of protection. The ozone resistance of both ENR-25 and ENR-50 may be improved by blending with ozone resistant rubber (e.g. a 50:20 blend of ENR-50 and EPDM, without antiozonant and strained to 20%, was crack-free after 14 days at 40°C and 50 pphm ozone).

7. APPLICATIONS

The properties of ENR suggest a wide range of potential applications, some of which are discussed below.

7.1. Tyres

7.1.1. Treads

The high wet traction of ENR (Table 10) obviously makes it attractive for tyre treads but other factors, such as rolling resistance and wear, need to be considered. Generally, with increasing hysteresis, the wet traction of rubbers increases, but this is at the expense of rolling resistance and hence fuel economy. With ENR-25 this pattern is broken, as at tyre operating temperatures the resilience of ENR-25 compounds, especially silica filled, is as high as or higher than that of an NR tread stock (Fig. 17).

Wet skid and rolling resistance measurements on tyres retreaded with ENR-25 compounds (Table 12) show that these tread stocks have both superior wet grip to an oil extended styrene butadiene rubber (OESBR 1712), and lower rolling resistance than NR.[24] These results are summarized in a Morton–Krol-type plot in Fig. 18. The tyre wear properties of ENR-25 black and black/silica (35:15) filled compounds are comparable with an NR tread stock. Improvements in wear can be obtained by blending with BR.

Although this reduces wet grip somewhat, black/silica (35:15) blends of ENR-25 with 20–30 phr of BR have rolling resistance comparable with that of NR and the wet grip of OESBR, combined with wear properties comparable with those of the latter.[28]

These attractive tread properties are currently being evaluated in both solid and pneumatic tyres.

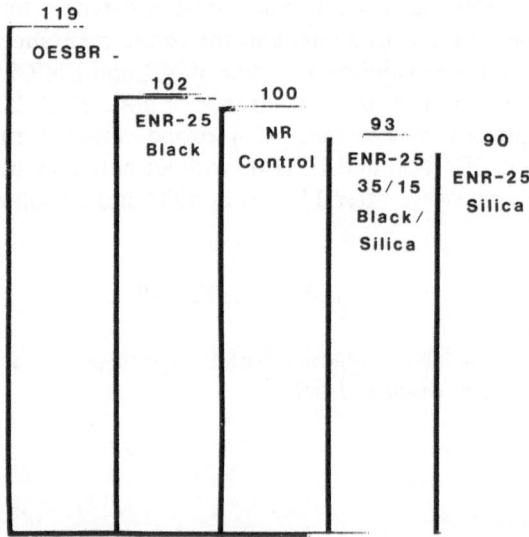

FIG. 17. Rolling resistance ratings of steel radial tyres retreaded with various rubbers.

TABLE 12
TYRE TREAD FORMULATIONS (phr)

	NR	OESBR	ENR-25	ENR-25	ENR-25
SMRL	100	—	—	—	—
OESBR 1712	—	100	—	—	—
ENR-25	—	—	100	100	100
Sodium carbonate	—	—	0·3	0·3	0·3
ISAF N220	50	50	45	35	—
Ultra-Sil VN3	—	—	—	15	50
Process oil	4	4	4	4	4
Antioxidant[a]	2	2	2	2	2
Zinc oxide	5	5	5	5	5
Stearic acid	2	2	2	2	2
Sulphur	1·5	1·8	1·4	2·4	2·4
MBS	1·5	0·7	1·4	1	1·2
CTP[b]	0·1	—	0·2	0·1	—
Hardness (IRHD)	67	67	67	65	64

[a] N-(1,3-Dimethylbutyl)-N'-phenyl-p-phenylenediamine.
[b] N-Cyclohexylthiophthalimide.

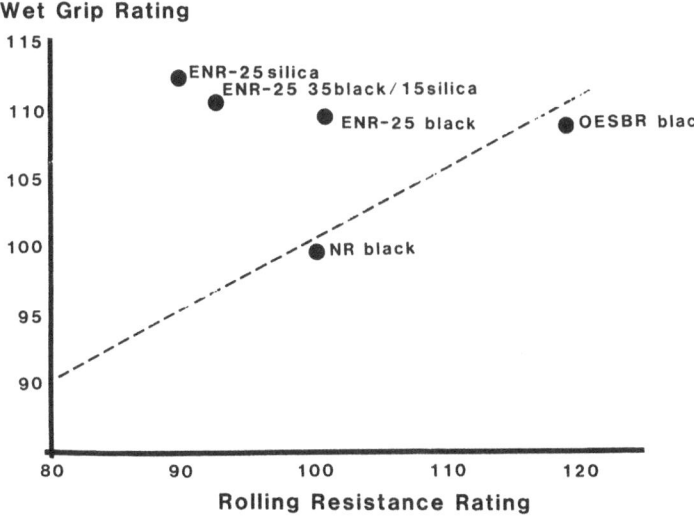

FIG. 18. Wet grip rating (pebble) versus rolling resistance for steel radial tyres retreaded with various rubbers.

7.1.2. Inner Liners and Tubes

The low air permeability properties of ENR-50 make it suitable for these applications. Although tack and cured adhesion values of the neat polymer to a tyre carcass stock are low, ENR-50/NR blends have higher adhesion values than those of Cl IIR/NR to a carcass stock. ENR-50/NR compounds also have faster cure rates, high tensile strength and elongation at break, higher tear resistance and better De Mattia flex and cut growth resistance.[29]

7.2. Damping

ENR-50 is a high-damping rubber at ambient temperatures by nature of its increased glass transition temperature (T_g). However, the relatively high T_g causes stiffening at a temperature around 0–5°C. Blending ENR-50 with other polymers and plasticizers increases its low-temperature service range. The dynamic properties of ENR-25 and an ENR-50/NR blend (Table 13) are recorded in Fig. 19. Both compounds have relatively high damping properties, do not show marked stiffening above −20°C and generally have better mechanical properties than blends based on acrylonitrile/isoprene copolymers (Kryrac 833).

TABLE 13
HIGH-DAMPING ENR FORMULATIONS (phr)

	A	B
NR (SMRL)	—	25
ENR-25	100	—
ENR-50	—	50
BR	—	25
Sodium carbonate	1·5	1·5
Calcium stearate	1·5	1·5
HAF (N330) black	45	45
Aluminium silicate	20	20
Adipate plasticizer	5	20
Aromatic process oil	25	20
Zinc oxide	5	5
Stearic acid	2	2
Antioxidant	3	3
Wax	3	3
Sulphur	0·4	0·4
MBS	2·5	2·5
TMTD	0·5	0·5

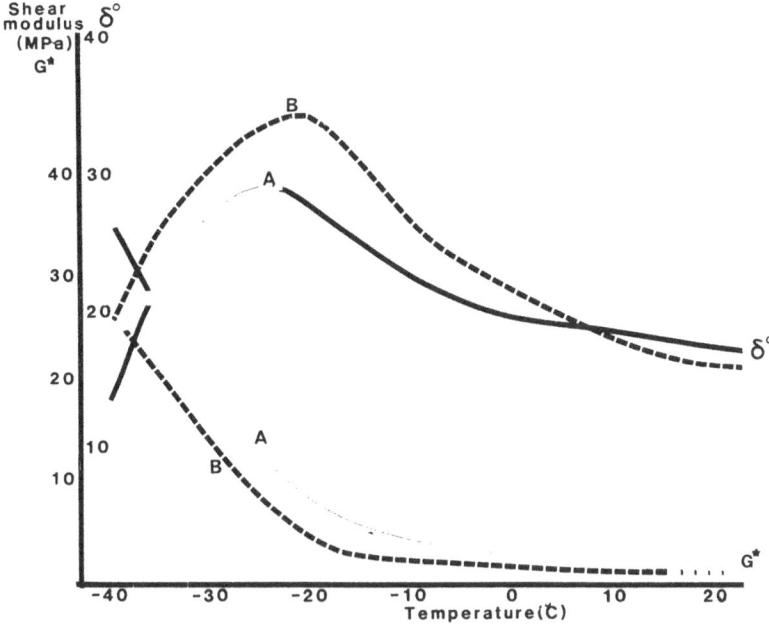

FIG. 19. Dynamic properties of ENR compounds (Table 13) at 1 Hz and 2·0% strain.

7.3. General Rubber Goods

A wide range of general rubber applications can be envisaged for ENR based on the properties described. ENR exhibits some unique properties which can be used to advantage. For example, unfilled vulcanizates, or vulcanizates filled with non-reinforcing fillers, are oil-resistant and have low gas permeability but exhibit the high strength properties associated with NR. Also black filled vulcanizates generally have higher tensile and fatigue properties than the speciality non-crystallizing rubbers such as NBR. Silica may be used to replace blacks where white, or coloured, products are desirable without having to use coupling agents.

Development work and service trials are being carried out in a wide range of rubber products, e.g. milking inflations, pharmaceutical and food products, oil hose and seals, conveyor belting, shoes soles and adhesives.

REFERENCES

1. KRAUS, G. and REYNOLDS, W. B., *J. Amer. Chem. Soc.*, **72**, 1950, 5621.
2. VAN VEERSEN, G. J., *Proc. 2nd Rubber Technol. Conf.*, London, 1948, 87.
3. VAN VEERSEN, G. J., *Rev. Trav. Chim.*, **69**, 1950, 1365.
4. ALLEN, P. W., BELL, C. L. M. and COCKBAIN, E. G., *Rubber Chem. Technol.*, **33**, 1960, 825.
5. HARDMAN, H. V. and HARDMAN, K. V., US Patent 2 349 549, 1944.
6. BARNARD, D., DAWES, K. and MENTE, P. G., *Proc. Int. Rubber Conf.*, Kuala Lumpur, **4**, 1974, 215.
7. CAIN, M. E., GAZELEY, K. F., GELLING, I. R. and LEWIS, P. M., *Rubber Chem. Technol.*, **45**, 1972, 204.
8. BAKER, C. S. L., BARNARD, D. and PORTER, M. R., *Rubber Chem. Technol.*, **43**, 1970, 501.
9. DAWES, K. and ROWLEY, R. J., *Plast. Rubber. Mater. Appl.*, **3**, 1978, 23.
10. GREENSPAN, F. P., in *Chemical Reactions of Polymers*, ed. E. M. Fettes, Interscience, New York, 1964.
11. SWERN, D., *Organic Peroxides*, Vol. 2, Wiley–Interscience, New York, 1971, Chapter 5.
12. ROSOWSKY, A., in *Heterocyclic Compounds with Three and Four Membered Rings*, Part 1, Interscience Publishers, New York, 1964.
13. SEFTON, M. V. and MERRILL, E. W., *J. Appl. Polym. Sci.*, **20**, 1976, 157.
14. NG, S. C. and GAN, L. H., *Europ. Polym. J.*, **17**, 1981, 1073.
15. GELLING, I. R., Malaysian Rubber Producer's Association, British Patent 2113692, 1985.
16. GELLING, I. R., *Rubber Chem. Technol.*, **58**, 1985, 86.
17. WITNAUER, L. P. and SWEN, D., *J. Amer. Chem. Soc.*, **72**, 1950, 3364.
18. JAY, R. R., *Analyt. Chem.*, **36**, 1964, 667.
19. SWERN, D., FINDLEY, T. W., BILLEN, C. N. and SCANLAN, J. T., *Analyt. Chem.*, **36**, 1947, 414.
20. DAVEY, J. E. and LOADMAN, M. J. R., *Brit. Polym. J.*, **16**, 1984, 134.
21. LEE, T. K. and PORTER, M., *Proc. Int. Rubber Conf.*, Venice, 1979, p. 991.
22. AMU, Abu bin, ISMAIL, Ku Abdul Rahman bin Ku, and DULNGALI, Sidek bin, *Int. Rubber Conf.*, Vol. II, Kuala Lumpur, 1985, p. 289.
23. GELLING, I. R. and SMITH, J. F., *Proc. Int. Rubber Conf.*, Venice, 1979, p. 140.
24. BAKER, C. S. L., GELLING, I. R. and NEWELL, R., *Rubber Chem. Technol.*, **58**, 1985, 67.
25. GELLING, I. R. and MORRISON, N. J., *Rubber Chem. Technol.*, **58**, 1985, 243.
26. DAVIES, C., WOLFE, S., GELLING, I. R. and THOMAS, A. G., *Polymer*, **24**, 1983, 107.

27. SAMSURI, Azemi bin, THOMAS, A. G. and GELLING, I. R., *Int. Rubber Conf.*, Vol. II, Kuala Lumpur, 1985, p. 386.
28. BAKER, C. S. L., GELLING, I. R. and PALMER, J., *Int. Rubber Conf.*, Vol. II, Kuala Lumpur, 1985, p. 336.
29. LOH, PANG CHAI and SEE TOH, MOK SANG, *Int. Rubber Conf.*, Vol. II, Kuala Lumpur, 1985, p. 312.

Chapter 4

PROCESS AIDS AND PLASTICIZERS

B. G. Crowther

Malaysian Rubber Producers' Research Association, Tun Abdul Razak Laboratory, Brickendonbury, Hertford, UK

1. INTRODUCTION

This chapter deals with materials used by the rubber industry to facilitate the processing of elastomers prior to the vulcanization stage. Much of the work discussed was derived from a number of examinations of specific materials carried out in the laboratories of the Malaysian Rubber Producers' Research Association (MRPRA). There is a dearth of such information in a published form; thus it is felt that a detailed discussion of the work undertaken would be beneficially published in a laboratory-style reporting form.

Due to the style of presentation many process aids have to be described by their trade names because they are of an undisclosed nature and therefore cannot be generalized. Natural rubber was used as the polymer in the investigations reported, which are part of a longer programme of research on processing being carried out at MRPRA to meet the demands of modern processing techniques. Many of the findings, however, will be applicable to other polymers.

2. DEFINITIONS AND DEVELOPMENT

2.1. Process Aid

The ISO definition of a process aid is a compounding material which improves the processability of a rubber compound. The ideal parameters for a process aid are that it

(1) reduces polymer nerve,
(2) aids better dispersion,
(3) reduces power consumption,
(4) promotes compound flow characteristics,
(5) does not affect finished properties, and
(6) acts at low dosage level.

Materials used for the purposes of process aids are:

(1) chemical peptizers;
(2) fatty acid soaps;
(3) fatty acid esters;
(4) petroleum oils;
(5) factice;
(6) resins;
(7) reclaim;
(8) natural rubber and partially vulcanized natural rubber;
(9) liquid polymers;
(10) waxes.

Process aids, especially fatty-acid based materials, have the ability to act in a beneficial way in situations which involve high shear rates, e.g. in extrusion or injection moulding.

2.2. Uses of Process Aids
The rubber industry as we know it today was founded upon natural rubber from the tree *Hevea brasiliensis*. The latex was tapped very crudely from wild trees growing in the jungles of Brazil. The latex was coagulated by pouring over a slowly revolving stick over a smoky fire, which caused drying and coagulation of the latex. The rubber from this source, known as Fine Para in its best grade, was a very hard, tough, high-viscosity material which must have caused considerable problems in the process we now call mastication. Although devices such as Hancock's 'Pickle' were used to make the rubber more amenable for further processing, there was soon an obvious need for compounding with fillers and vulcanization ingredients. Various natural oils, resins, pitches, blown bitumen (known as mineral rubber), fats and greases from animal sources were all investigated in those early days as possible desirable ingredients for rubber formulations. In turn the

petroleum oil industry proved to be another source of plasticizers and eventually offered ranges of oils for specific applications and polymers; in doing so it found the rubber industry a useful, if small, outlet for some of its lower grade materials such as very heavy aromatic oils.

Some of these early materials found acceptability, but often had to be discarded when it was found that they were incompatible with the rubber or caused staining of light-coloured compounds or migration stains to other materials coming into contact with the vulcanized rubber component containing them.

2.3. Synthetic Rubber Development

The advent, from the early 1930s onwards, of synthetic rubbers with processing characteristics quite different from natural rubber created new demands on both equipment and compounding ingredients.

The early commercial synthetic rubbers created were the polysulphide rubbers now known as Thiokols: these rubbers had good oil resistance, unlike natural rubber, but with a very serious drawback which prevented their widespread acceptance, namely their smell. Processing of these materials in factory-size batches proved to be very obnoxious. Acrylonitrile–butadiene copolymers (nitrile rubbers) were developed by Bayer and were to find eventually worldwide acceptance in many applications requiring oil resistance. In the mid 1930s polychloroprene rubbers became a commercial proposition after early experimental work at Du Pont. These rubbers, intermediate in properties between nitrile and natural rubber, soon found wide acceptance.

With the onset of World War II, immediate work was started to find a commercial substitute for natural rubber because of the restrictions on world movement of goods and likelihood of the loss of rubber-growing areas in the Far East to the Japanese. Some commercial quantities of GRS, a styrene–butadiene copolymer, became available; and modern forms of this make up the rubber type in most popular use, especially in car tyres. After World War II renewed interest in the field of fundamental polymerization technology soon produced a number of new synthetic rubbers, many of which required special additives to assist their processing, whilst others such as ethylene–propylene terpolymer allowed very large volumes of petroleum oil to be used as extenders rather than in limited quantities as plasticizers. Work in the field of natural rubber also enabled large volumes of oil to be used with this polymer.

2.4. How Process Aids Work

Fatty-acid based process aids can reduce the viscosity of a rubber containing them, without an actual chain scission and reduced molecular chain length being involved. The viscosity reduction is believed to be achieved by simple lubrication of the molecular entanglement, allowing individual chains to slip easily relative to their fellow chains and thus creating the illusion of the viscosity reduction normally associated with short-chainlength molecular movement.

For the purposes of this paper, plasticizers (which are usually associated with liquid additives, e.g. petroleum oils) are specific types of process aids used in unvulcanized rubbers to reduce viscosity. Other plasticizers, for example of the ester type, may be used in synthetic rubbers for other reasons such as the lowering of the glass transition point for low-temperature applications.

Other materials, sometimes known as chemical plasticizers and more commonly as peptizers, fulfil the role of reducing the viscosity of the polymer to which they are added during the chain scission stage of mastication, and act as short-stops to prevent the molecular chains broken by shear forces recombining into long-chain molecules again.

Another class of aid is that which assists with filler dispersion, increase of tack of compounds and the improvement of dimensional stability of extrudates (see Section 5).

3. PLASTICIZERS FOR NATURAL RUBBER

The generally recognized plasticizers or softeners for natural rubber are petroleum oils. These oils are available in three categories (aromatic, paraffinic and naphthenic) and are the products which are derived from crude oil after the more volatile petrol and heating oil fractions have been removed by distillation; consequently they are materials which exhibit a wide distribution of molecular weight.

3.1. General Characteristics -

Although the oils are sold as three types, they are always blends of all three hydrocarbon forms but having a predominance of one of them (although this need not be the case with 'naphthenic' types). In the case of 'paraffinics', for instance, the only true paraffins present are waxes, whilst there is a higher preponderance of molecules having paraffinic side chains attached to them. (For further discussion of oils and their constituents, see ref. 1.)

In general terms, oils of an aromatic preponderance give best processability in general-purpose rubbers but are most prone to causing stains and poor colour, and may contribute to poorer ageing resistance. Paraffinic oils are less effective in their assistance to processing but have little effect upon ageing, contact staining or colour stability. Naphthenic oils give processing and behaviour characteristics in natural rubber between those of aromatics and paraffinics.

3.2. Terms Used

3.2.1. Viscosity Gravity Constant

Table 1 shows a classification used by BP Oil Ltd[4] of the oil types in use in the rubber industry. This classification can be summarized into three main types.

(1) Paraffinic oils > 50% C_P, VGC < 0·850.
(2) Naphthenic oils 30–40% C_N, VGC 0·850–0·900.
(3) Aromatic oils > 35% C_A, VGC > 0·950.

The term viscosity gravity constant (VGC) is an expression of the overall aromaticity of the oil, independent of molecular weight. C_P, C_N, C_A are respectively paraffinic, naphthenic and aromatic carbon entities. VGC appears as the constant (a) in the following expression relating oil viscosity to specific gravity:

$$G = a\frac{1·0752 - a}{10}\log(V + 38)$$

TABLE 1

CLASSIFICATION OF RUBBER PROCESS OILS[a]

	Oil type						
	Paraffinic	Relatively paraffinic	Naphthenic	Relatively aromatic	Aromatic	Very aromatic	Extremely aromatic
VGC range	0·790–0·819	0·820–0·849	0·850–0·899	0·900–0·949	0·950–0·999	1·00–1·05	>1·05
C_A(%)	0–10	0–15	10–30	25–40	35–50	50–60	>60
C_N(%)	20–35	25–40	30–45	20–45	20–40	0–25	<25
C_P(%)	60–74	50–65	35–45	25–45	20–35	0–25	<25

[a] Reproduced by courtesy of BP Oil Ltd, UK.

where G is specific gravity at 15·5°C and V is the Saybolt viscosity at 100°C.

3.2.2. Aniline Point

The aniline point of an oil is derived from the temperature at which equal volumes of oil and aniline are mutually soluble. Although this term has now to a large degree been superseded by VGC, it is still used by the rubber industry in describing and specifying oils.

The higher the aromaticity of the oil the lower the aniline point. The aniline point can be influenced by the molecular weight of the oil and can be difficult to determine when using very dark and opaque oils.

3.2.3. Flash Point

This parameter has sometimes been used as a guide to the amount of oil to be expected to be left during high-temperature service of components containing the oil; other factors such as polymer and filler compatibility with the oil have a bearing on such losses.

3.3. Effects of Oils in a Test Compound

To illustrate the effect of a number of commercial oils on the properties of both unvulcanized and vulcanized natural rubber compounds, two levels of oil were applied to a standard truck-tyre type formulation. The oils in the evaluation are listed in Table 2 and the compound used in Table 3. The range of oils examined covers much of the range of viscosities normally used in industry. For ease of reference in subsequent tables the oils are referred to by the reference number indicated in Table 2.

3.3.1. Mooney Viscometer and ODR Results

At an oil level of 10 phr the Mooney viscosity was generally lower with the naphthenics, but at 20 phr the paraffinic oils are found to be slightly better than the aromatics for viscosity reduction.

Figures 1 and 2 show the Mooney viscosity values obtained from a single Banbury BR batch at each oil level.

Table 4 gives the various parameters derived from a Monsanto Rheometer examination of the mixes at 150°C. In general, little difference was noted among the oils at both levels.

3.3.2. Vulcanizate Properties

A vulcanization time of 20 min at 150°C ensured that the mixes were at or slightly beyond optimum cure.

TABLE 2
PROCESS OILS USED

Ref. no.	Oil	Saybolt viscosity		VGC
		38°C	100°C	
	Aromatic			
1	Dutrex 729[a]	4360	210	0·991
2	Sundex 795[b]	3500	96	0·954
3	Enerflex 84[c]	—	81	0·981
	Naphthenic			
4	Circosol 4240[b]	2520	85·9	0·883
5	Sunthene 4240[b]	2206	84·7	0·882
6	Sunthene 380[b]	760	60	0·869
7	Circolight Process Oil[b]	156	41·0	0·878
8	Fina 2059[b]	—	43·5	0·830
9	Circosol 410[b]	110	38·2	0·886
10	Sunthene 410[b]	104	38·0	0·871
	Paraffinic			
11	Sunpar 2280[b]	2642	155	0·800
12	Sunpar 150[b]	500	63·5	0·803
13	Fina 2069[d]	—	42·3	0·805
14	Sunpar 110[b]	110	40·4	0·807
15	Enerpar 20[c]	—	38·0	0·832

[a] Shell Oil Ltd.
[b] Sun Oil Ltd. (Sun Oil has ceased trading in the UK. A similar range of oils will now be marketed by Schill & Seilacher under the names Strukpar (paraffinic), Struksol (naphthenic), Strukthene (naphthenic) and Strudex (aromatic).)
[c] BP Oil Ltd.
[d] Petrofina.

3.3.2.1. *Hardness* (see Fig. 3). Although there were slight differences among the individual oils used, presumably caused by differences in oil viscosity, there is little effect of class of oil at 10 phr. At 20 phr the paraffinic oils had a slightly greater softening effect upon the rubber compound than the other two types (see Table 5). There does not seem to be a relationship between Saybolt viscosity and hardness.

3.3.2.2. *Tensile strength* (see Fig. 4). At both levels of addition there was no effect of the different classes of oils or of Saybolt viscosity on unaged tensile strength. No significant effect of type or viscosity of

TABLE 3
FORMULATION FOR OIL EVALUATION

	Parts by weight
SMR 10[a]	100
Zinc oxide	5
Stearic acid	2·5
N-330 HAF black	45
N-Isopropyl-N'-phenyl-p-phenylenediamine	1·5
Oil	10 or 20
N-Oxydiethylenebenzothiazole-2-sulphenamide	0·8
Sulphur	2·5

MIX CYCLE: BR BANBURY SPEED II

Time (min)	Action
0	Add polymer.
0·5	Add small powders.
1·0	Add half black.
2·0	Half black plus oil.
3·5	Dump.

[a] Standard Malaysian Rubber.

oil on tensile strength after ageing at 70°C and 100°C was observed but individual oils did give widely differing values. The addition of 20 phr of oil gave less variation in aged tensile strength values.

3.3.2.3. Compression set (see Fig. 5). Compression set measurements were carried out after 1 day at 100°C or 7 days at 70°C under 25% compression strain. At the 10 phr level little effect was noticed at either condition of test. At 20 phr there appeared to be a slight advantage in the use of paraffinic or naphthenic oil rather than an aromatic one.

3.3.2.4. Dunlop resilience (see Fig. 6). In terms of Dunlop resilience, differences between oil types were noticeable. Aromatic oils generally gave lower resilience, particularly at the higher dosage. Paraffinic oils generally had higher resilience, with values at 20 phr some 10% above those for aromatic oils. The effect was also related to

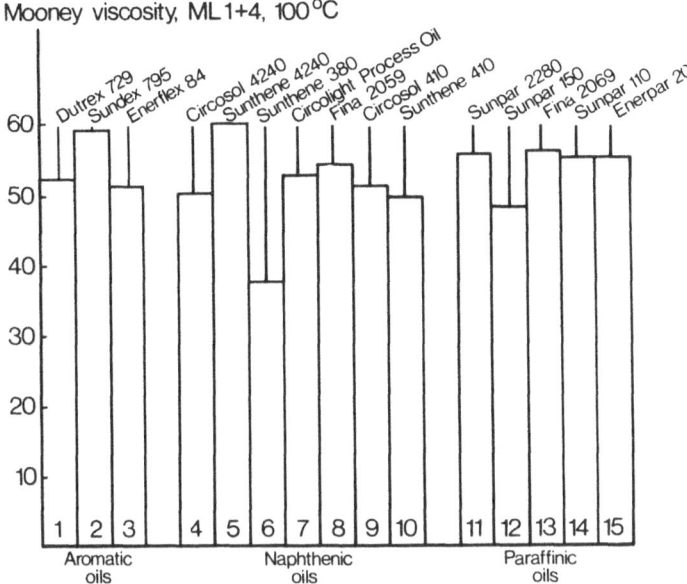

FIG. 1. Mooney viscosity, ML1 + 4, 100°C, at 10 phr of oil. The oils are listed in the order shown here in Figs 2–6.

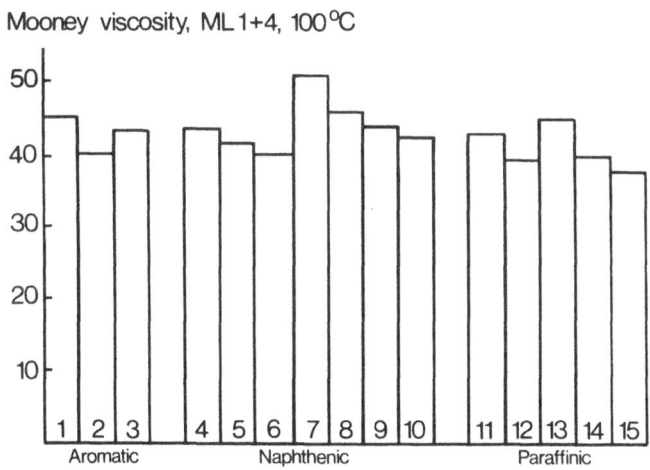

FIG. 2. Mooney viscosity, ML1 + 4, 100°C, at 20 phr of oil.

TABLE 4
RHEOMETER DATA

Oil	Type[a]	10 phr oil				20 phr oil			
		t_{s5} (min)	$t_c(90)$ (min)	M_{HR} (torque units)	M_L	t_{s5} (min)	$t_c(90)$ (min)	M_{HR} (torque units)	M_L
Dutrex 729	A	4·12	14·33	33·0	7·35	4·51	16·00	24·3	8·46
Sundex 795	A	3·42	13·13	35·6	7·45	3·58	14·32	27·2	7·08
Enerflex 84	A	4·16	13·47	32·9	8·25	4·25	14·28	30·3	9·00
Circosol 4240	N	3·58	14·31	32·5	7·39	4·39	15·18	28·7	8·16
Sunthene 4240	N	3·53	13·40	35·5	7·36	4·28	15·11	27·4	8·25
Sunthene 380	N	5·16	14·51	30·2	9·25	5·00	16·05	25·4	8·53
Circolight PO	N	3·52	14·43	33·5	7·41	4·17	14·50	30·5	7·50
Fina 2059	N	3·47	14·28	32·1	8·16	4·25	15·24	27·8	8·42
Circosol 410	N	3·48	14·21	35·5	7·37	4·17	16·02	27·2	8·22
Sunthene 410	N	4·15	13·56	32·7	7·58	4·31	15·18	27·6	7·58
Sunpar 2280	P	4·07	14·20	34·6	7·44	4·26	15·13	28·0	8·07
Sunpar 150	P	3·57	14·22	33·4	7·52	4·35	14·32	28·8	8·16
Fina 2069	P	4·04	14·45	33·9	7·57	4·35	16·28	26·3	8·56
Sunpar 110	P	4·00	13·45	32·5	7·13	4·29	14·58	27·2	8·28
Enerpar 20	P	4·21	14·06	35·3	7·28	4·39	15·00	26·4	8·14

[a] A, Aromatic; N, naphthenic; P, paraffinic.

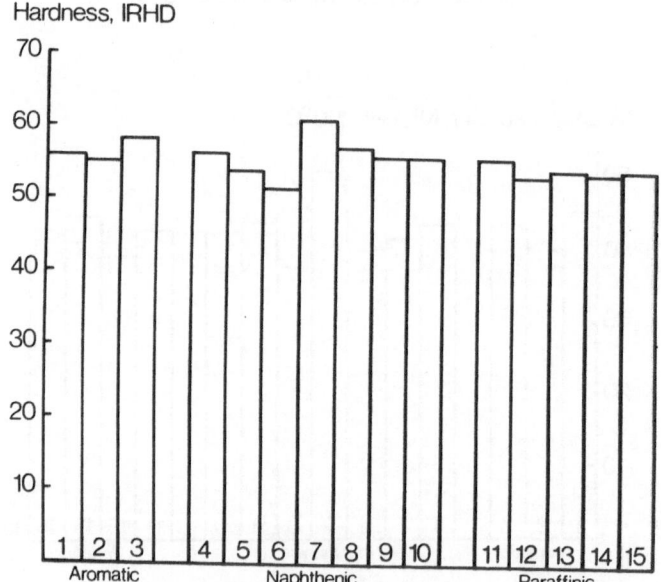

FIG. 3. Hardness, IRHD, at 20 phr of oil.

TABLE 5
AVERAGE VALUES OF HARDNESS FOR THE
THREE OIL TYPES

Dosage (phr)	A	N	P
10	62	61·5	62
20	56·5	56	54·5

viscosity, the highest viscosity oils giving lower resilience at the 20 phr level.

3.4. Discussion of Results

Such an evaluation of oils from different suppliers is rarely seen in print and such full detail has been presented to give compounders some idea of the potential features of their own particular types or viscosities of oil. It is obvious that although in most compounds few significant differences in properties will be noticed certain parameters concerned with highly stringent specifications may be influenced either

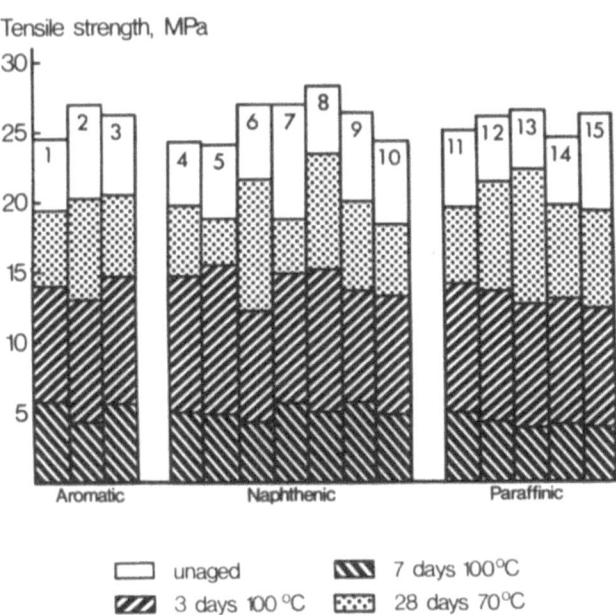

FIG. 4. Aged and unaged tensile strength at 20 phr of oil.

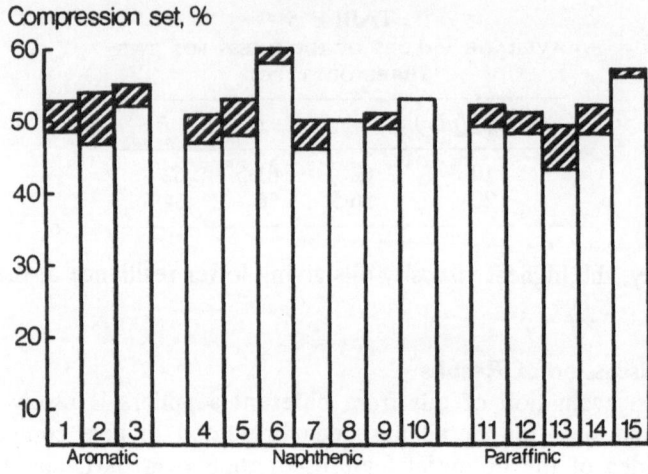

FIG. 5. Compression set at 20 phr of oil.

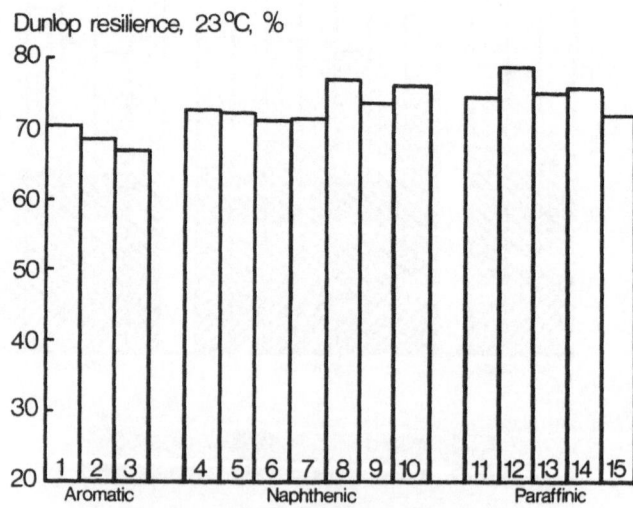

FIG. 6. Dunlop resilience at 20 phr of oil.

by the class of oil used, or indeed by the viscosity of the oil in common factory use. In many cases, of course, other factors mentioned earlier such as migration staining and influence on cure characteristics, especially by aromatic oils, may, together with cost parameters, dictate the oil to be used in a particular application.

4. PEPTIZERS AND VISCOSITY REDUCERS

Although petroleum oils reduce compound viscosity, process aids are now available that are more effective in this respect at much lower levels of addition, so minimizing the effect on vulcanizate properties. These materials are as follows:

(1) chemical peptizers;
(2) fatty-acid soaps;
(3) blends of chemical peptizers and fatty-acid soaps.

These types of material may be termed viscosity reducers although their mode of working is very different in nature.

4.1. Why Viscosity Reducers are Necessary
Naturally occurring polyisoprene, i.e. natural rubber, has one characteristic which makes it unlike most synthetic polymers; it has a high viscosity, usually acquired during shipment and storage. (Viscosity-stabilized natural rubber is now readily available from Malaysia, marketed as SMR LV and SMR CV, with three distinct controlled viscosity ranges, and SMR GP. The viscosity control is obtained by the incorporation, during production, of hydroxylamine salts, which act as inhibitors of viscosity increase.) The high initial viscosity of most grades of natural rubber is not suitable for the majority of factory processes. Considerable energy has to be expended by the consumer factory in reducing the viscosity to the level required, either as a separate mastication stage, or by a longer mixing cycle which incorporates polymer breakdown as the initial step in the cycle. Materials which accelerate the breakdown of the natural rubber and reduce energy consumption are therefore of vital interest to industry.

4.2. Chemical Peptizers
The most common chemical peptizers in general use are based on a number of chemical classes. These are aromatic mercaptans, arylamines, sulphonic acids and derivatives, pentachlorothiophenol and

dithiobisbenzamide. These materials may be used alone or in combination with various activators or dispersing aids.

4.2.1. Level of Use

Peptizers have the effect of accelerating oxidative radical chain scission catalytically and enable this mechanism to take place at lower temperatures than would be the case in unassisted mastication. Mastication at lower temperatures without peptizers is effective through shear-induced chain scission but as the temperature rises and shearing forces fall the process becomes less efficient until at about 120°C it is prohibitively slow. At higher temperatures the oxidative mechanism becomes operative. To achieve the catalysis of this mechanism only very small proportions of the peptizer are required, often as low as 0·05 phr and usually not greater than 0·3 phr. It is obvious that such small amounts of chemicals must be very efficiently dispersed quickly throughout the mass of the rubber if a uniform reduction of viscosity is to be achieved. This is often not the case with pure peptizers, resulting in pockets of very low-viscosity (honey-like consistency) material being produced. Subsequent mixing does, of course, disperse such low-viscosity material, but does detract to some extent from the physical behaviour and handling of a batch of peptized rubber, because of surface stickiness.

There is also the modern mixing trend to take into account, whereby the breakdown of the rubber takes place as an integral part of the batch mixing cycle. This is a relatively inefficient way of peptizing rubber because of the lower shear conditions in the mixing chamber owing to a small volume of rubber being used relative to chamber size to allow for addition of the fillers as the second part of the mixing cycle. It has been found necessary in this instance to increase the level of peptizer used by 100% to achieve the same viscosity drop that one would obtain if using a full mixer load of raw polymer.

4.2.2. Processing Trials

The processing of natural rubber with chemical peptizers both in single-stage, full-chamber mastication and as part of a mixing cycle (called 'in batch' peptization) was investigated for efficiency of breakdown and possible effects of chemical peptizers on subsequent vulcanizate properties.

4.2.2.1. Equipment and materials used. The machine used for this work was a K2A Shaw Intermix of batch volume 26 litres, which is believed to give a performance reasonably comparable with that in rather larger-scale production units. Processing was carried out using a range of rotor speeds and ram pressures to simulate the variety of mixing conditions used in industry.

In most of the work, the rubber used was SMR 10 and Renacit VII (pentachlorothiophenol; Bayer AG) was the main peptizer system, but some assessment of Pepton 44 (activated dibenzamidodiphenyl disulphide; Anchor Chemical Co. Ltd) was carried out for comparative purposes. Two levels, 0·1 and 0·15 phr, of peptizer, a range of Intermix rotor speeds and ram pressures and a range of cycle times were used. Subsequently Renacit VII and Pepton 44 were compared over a wide range of dosages using fixed Intermix settings and a standard 5-min processing time.

4.2.2.2. Mastication and peptization. To ensure uniformity of the base polymer stock, 15 33·3-kg bales of SMR 10 were each chopped into ten equal pieces and the pieces randomly mixed. Cold rubber was added to the mixer and masticated or peptized for the times shown in Fig. 7. The intermix conditions used are shown in Table 6.

The peptizer was placed in a small polyethylene bag which was inserted in the mixer after half the polymer had been added. This technique ensured that the peptizer was introduced to the chamber of the mixer and not merely deposited over the walls of the chute and the underside and faces of the ram, where it is ineffective.

After dumping, batches were milled on a 42-in mill for 3 min and six samples were taken from different areas of each batch. Mooney viscosity was measured on each sample after a 24 h rest.

The remaining stock pieces, from which the Mooney viscosity samples had been taken, were blended together and mixed in a laboratory BR Banbury using the formulation detailed in Table 7. The cure behaviour of each batch was examined using a Monsanto Rheometer and test pieces for assessing vulcanizate properties were cured for 20 min at 150°C. Tensile properties were measured after ageing in an air oven.

4.2.2.3. Increased levels of peptizers. The effect of using increased levels of peptizer (Fig. 8) was examined using Renacit VII at levels of 0·05–0·6 phr and Pepton 44 at levels of 0·05–0·5 phr. The SMR 10 blend described earlier was used with Intermix conditions as shown in Table 6, under D (low rotor speed and low ram pressure) and a

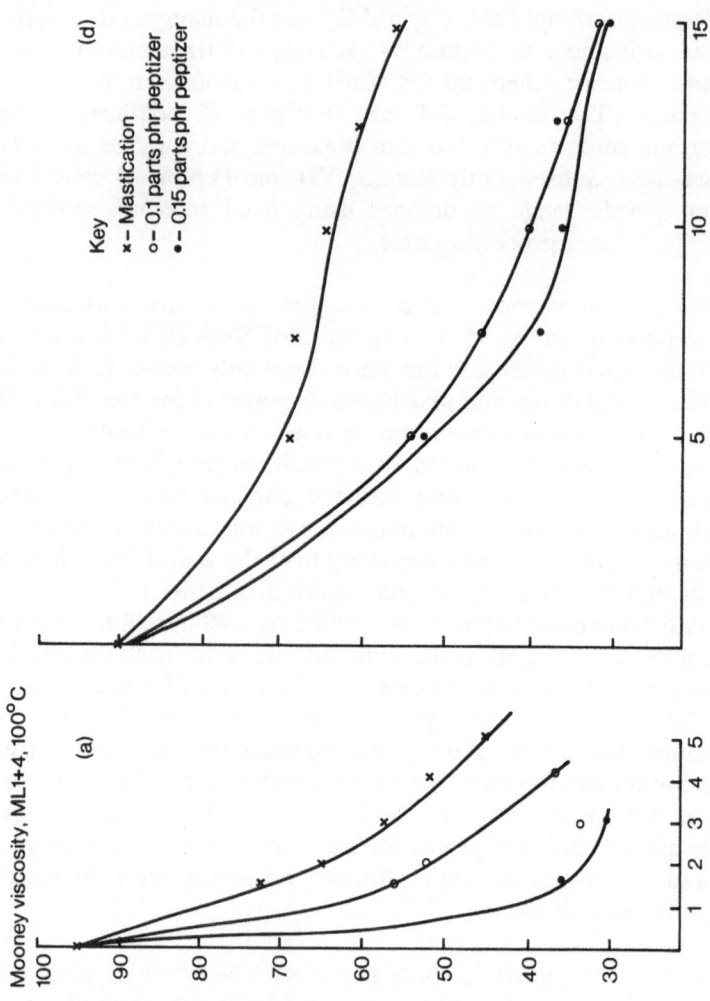

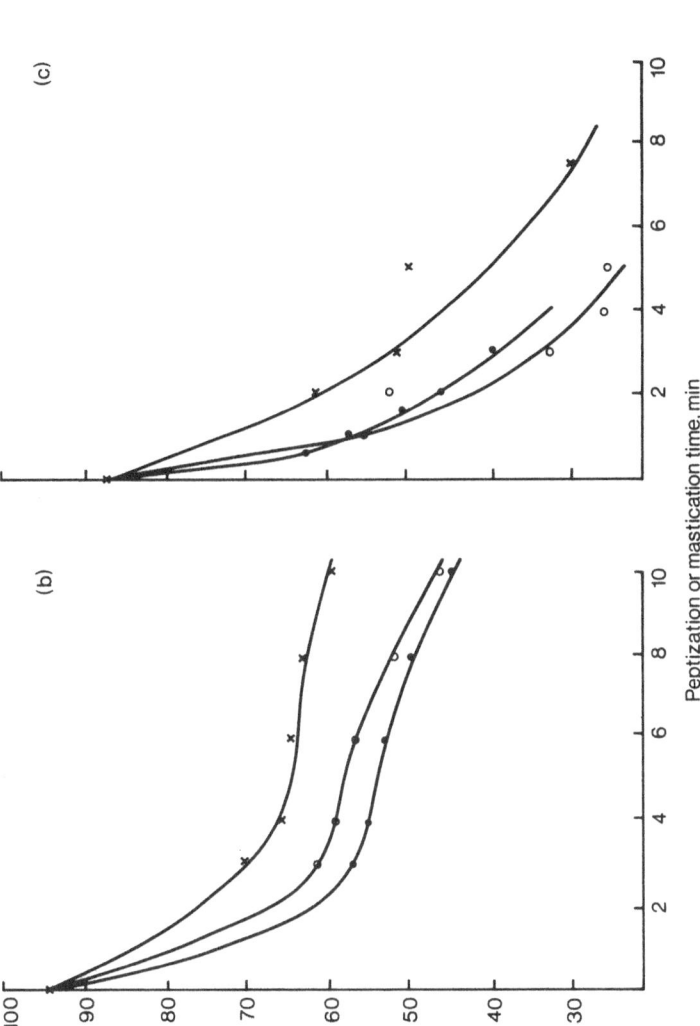

FIG 7. Viscosity versus mixing time for mastication and peptization. (a) High rotor speed, high ram pressure; (b) low rotor speed, high ram pressure; (c) high rotor speed, low ram pressure; (d) low rotor speed, low ram pressure.

TABLE 6

K2A INTERMIX CONDITIONS

	A	B	C	D
Chamber temperature (°C)	80	80	80	80
Rotor speed (rev/min)	50	25	50	25
Ram pressure (psi)	100	100	50	50
Water throughput (litre/h)	9000	9000	9000	9000
Charge of rubber (kg)	22	22	22	22

TABLE 7

FORMULATION FOR PEPTIZER EVALUATION

	Parts by weight
Masticated or peptized rubber (SMR 10)	100
Process oil[a]	5
Stearic acid	2·5
Zinc oxide	5
N-375, HAF black	45
IPPD[b]	1·5
OBS[c]	0·8
Sulphur	2·5

[a] Dutrex 729 (Shell Chem. Co.).
[b] N-Isopropyl-N'-phenyl-p-phenylenediamine.
[c] N-Oxydiethylenebenzothiazole-2-sulphenamide.

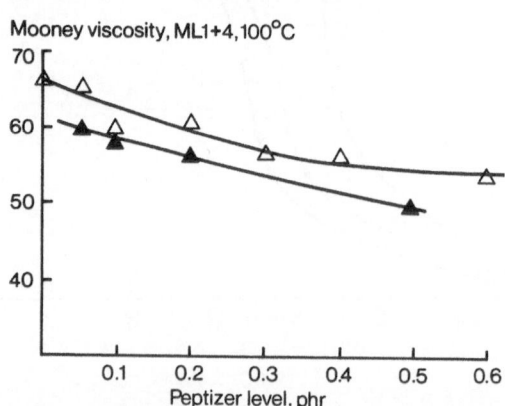

FIG. 8. Viscosity versus peptizer level for low-rotor-speed, low-ram-pressure mixing conditions, △, Renacit VII; ▲, Pepton 44.

peptization time of 5 min, followed by mill blending for 3 min. Mixes were processed into vulcanizates using the same formulations and conditions as before.

4.2.2.4. Peptization 'in batch'. If a separate peptization stage can be eliminated, savings in energy consumption and labour can be made. For peptization as part of a normal mix cycle to be effective, the peptizer needs to be added to the rubber and mixed for a short time, before the addition of fillers and other ingredients. This possibility was investigated using the Intermix under the conditions shown in Table 6 under A (high rotor speed and high ram pressure) and with the cycle shown below.

Time (min)	Action
0	Add rubber and peptizers.
1·5	Add carbon black, oil and powders (except accelerators and sulphur).
3	Dump.

For this work SMR 20 was used with the two peptizers at levels of 0·1–0·5 phr. Mixes were processed to vulcanizates using the same formulation and conditions as before.

4.2.2.5. Viscosity reduction results. Figure 7 shows the viscosity reductions achieved for mastication and peptization with 0·1 and · 0·15 phr of peptizer (Renacit VII) under the various machine conditions. Although, as can be seen, high-speed peptization gives large viscosity reduction in very short times, severe processing problems were encountered with some of the batches. At high rotor speed and high ram pressure (A) and with 0·1 phr of peptizer the batch mixed for 5 min was extremely sticky and under the same conditions with 0·15 phr of peptizer, batches mixed for more than 3 min were impossible to handle on the sheeting mill and had to be discarded. Although the remaining batches with 0·15 phr peptizer could be processed further under laboratory conditions it should be noted that under factory conditions such mixes would also have to be discarded. At high speed and low ram pressure (C), similar but less marked stickiness occurred and again batches with 0·1 phr of peptizer could only be mixed for 5 min and those with 0·15 phr only for 3 min.

The inconsistency of the batches mixed under high-speed, high ram pressure conditions can be seen clearly from Table 8, which lists the standard deviations for the six viscosity measurements made of each level. The high tack and nervy appearance of these batches are probably a result of inadequate dispersion of the peptizer before the start of chemical reaction and chain scission. This would create a blend of high and low molecular weight polymer which would then be very difficult to process.

Higher levels of both peptizers were used to assess whether more peptizer would give more useful viscosity reduction. As shown in Fig. 8 both peptizers gave a progressive drop in viscosity as the dosage increased but the effect was not linear, especially with Renacit VII, and above 0·3 phr of peptizer further additions gave very little benefit. Therefore the most appropriate level to use seems to lie between 0·05 and 0·15 phr. Under this condition Pepton 44 gave higher viscosity reductions than Renacit VII but after further processing the viscosities of the final mix were similar.

The effects of adding peptizer to SMR 20 during the initial polymer breakdown phase of an internal mixer cycle are shown in Fig. 9. Peptizers are deactivated by certain compounding ingredients, particularly carbon black; so it is important that the cycle allows sufficient time for the peptizer to act, before the remaining ingredients are added. With only the rubber present, the mixer is underloaded, and it was found that a relatively large amount of peptizer (ca 0·3 phr) was necessary to give useful reduction in viscosity.

TABLE 8

STANDARD   DEVIATION   VALUES   FOR   VISCOSITY   MEASUREMENTS
(INTERMIX AT HIGH SPEED, HIGH RAM PRESSURE)

	Standard deviation of six viscosity values				
	1·5 min	2·0 min	3·0 min	4·0 min	5·0 min
Mastication	1·17	0·49	0·41	0·49	0·58
0·1 phr chemical peptizer	3·94	7·67	9·14	2·11	5·5
0·15 phr chemical peptizer	8·12	6·69	13·3	—	—
Struktol A60 2 phr	—	—	0·71	—	0·89
Aktiplast 2 phr	—	—	0·45	—	—
Zinc stearate 2 phr	—	—	0·91	—	—

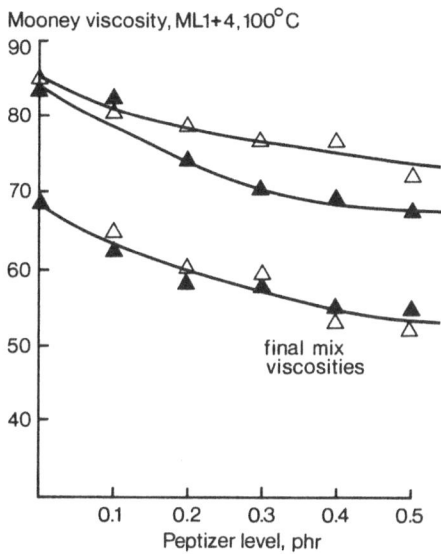

FIG. 9. Viscosity versus peptizer level for 'in batch' peptization using a short, high-rotor-speed, high-ram-pressure mixing cycle. Symbols as in Fig. 8. The top two curves represent the viscosity of the masterbatch stage in each case with the two peptizers.

4.2.2.6. Vulcanizate properties. Figure 10 shows the variation of tensile strength using the different Intermix conditions. Since there was no simple dependence of properties on time of mastication or peptization the values given were the means over a range of mastication or peptization times: thus the tensile strength value given for mastication under condition A is the mean value over five batches masticated for 1, 2, 3, 4 and 5 min. The peptized batches show very similar performance to the masticated ones and in some cases, e.g. at low speed and low ram pressure (condition D), the peptized batches gave slightly better unaged and aged tensile strength than the masticated batches. Generally, vulcanizate properties seem to show more dependence on the conditions of preparation of the batches than on the presence of peptizer.

Figure 11 shows the tensile strength values achieved from batches peptized using a range of Renacit and Pepton levels, both as a separate stage, and 'in batch'. Although there is some scatter in the results there is no evidence that either of the peptizers has a significant

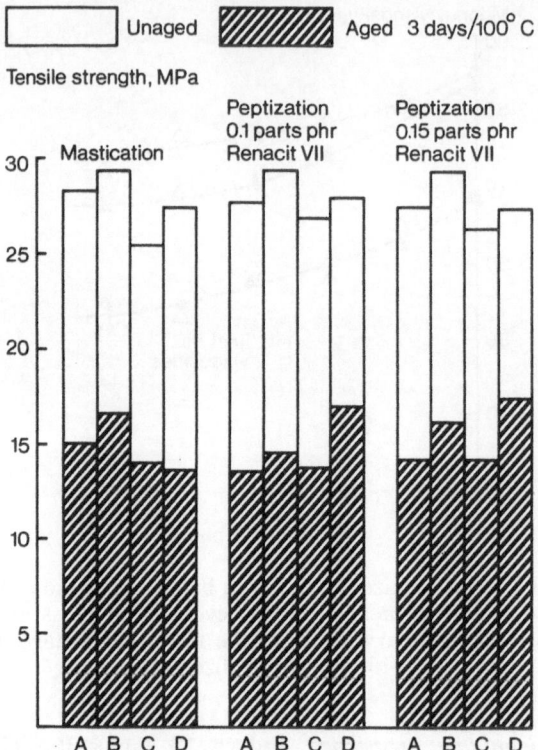

Fig. 10. Unaged and aged tensile strength values for vulcanizates prepared
from masticated and peptized rubbers.

	Rotor speed	Ram pressure
A	High	High
B	Low	High
C	High	Low
D	Low	Low

adverse effect on aged or unaged tensile strength, and similar results
were found for elongation at break and modulus at 200% elongation.

4.2.3. Usefulness of Chemical Peptization
The work indicates that peptization of natural rubber in modern
internal mixers poses problems not generally found with peptization

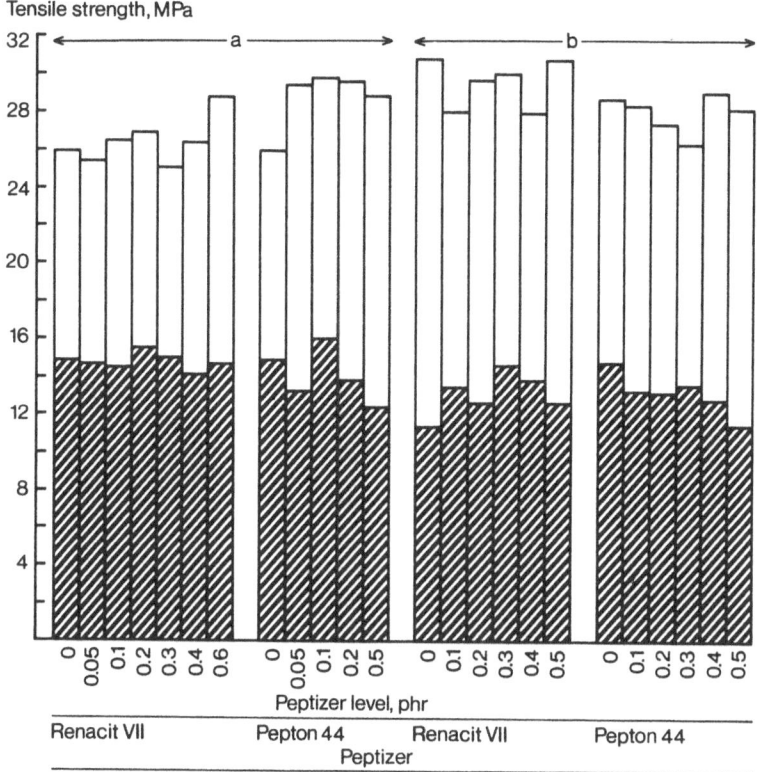

Fig. 11. Effect of different levels of peptizers on unaged and aged tensile strength. (a) Batches peptized for 5 min at low rotor speed and low ram pressure; (b) 'in batch' peptization. □, Unaged; ▨, aged.

using the open mills for which peptizers were originally designed. At high speed in the internal mixer, peptization resulted in an in-homogeneous product which exhibited extreme surface stickiness and was very difficult, in some instances impossible, to sheet on a sheeting mill. It seems likely that under these high-speed conditions, where temperatures can reach 150°C in under 2 min, the peptizer does not have sufficient time to disperse before it commences its action. Thus the polymer in the immediate vicinity of the peptizer breaks down very rapidly to form a 'syrupy' low molecular weight fraction, which then acts as a lubricant for the rest of the batch, preventing further breakdown of the remaining high molecular weight portion of the mass. Using low rotor speeds, such problems were not encountered and a steady viscosity reduction with time of peptization was obtained. At low speed, however, peptization gave similar viscosity reductions, comparable mixer times and similar energy consumption to unassisted mastication at high speed.

Peptization as part of a normal internal mixer mixing cycle did not give the stickiness associated with peptization as a separate stage, although of course low molecular weight material may well have been generated but 'absorbed' by the carbon black and thus not obvious. It is important to design the mix cycle to give sufficient time for the peptizer to act before other ingredients are added. A mixing time of 1·0–1·5 min at high speed, before carbon black is added, is appropriate. It was also found necessary to add higher levels of peptizer than those normally required. As shown here, for natural rubber peptizer levels of 0·05–0·15 phr are usually sufficient; higher levels give little further viscosity reduction. For in-batch peptization, however, levels of 0·3–0·5 phr are needed.

4.3. Fatty-Acid Soaps as Viscosity Reducers

Alternative 'peptizers', or viscosity reducers as they may more properly be called, are based upon fatty-acid soaps. These are known to be less potent than true chemical peptizers and as they are used at substantially greater concentrations, they may be more evenly and quickly distributed. Two types of material are available in this category: (a) fatty-acid soaps and (b) blends of such soaps with chemical peptizers. For convenience these are dealt with separately.

4.3.1. Materials Examined

The materials examined and representing classes available from a number of suppliers were as follows.

A　Aktiplast (Rhein Chemie). Blend of zinc soaps of higher molecular weight unsaturated fatty acids.

B　Dispergum T (Deutsche Oelfabrik GmbH—DOG). Zinc salts of defined unsaturated fatty acids with lubricants.

C　Struktol A60 (Schill & Seilacher). Mixture of zinc soaps of high molecular weight fatty acids.

D　Zinc stearate (used as comparison).

E　Dispergum 24 (DOG). Detailed composition not disclosed.

F Struktol A82 (Schill & Seilacher). Detailed composition not disclosed.

G Struktol A60 with Renacit VII 0·15 phr (see Section 6).

H Renacit VII.

I Dutrex 729 (see Table 2).

These materials were examined over a range of loadings to levels in excess of those recommended by their suppliers.

As a comparison, viscosity reduction was also achieved by carrying out mastication in the presence of a process oil, and, where appropriate, comparative data have been included. It should be noted, however, that considerable difficulty was encountered when adding the larger volumes of oil. Mixing cycles, which started at 3 min for the lower oil loading, had to be extended by up to 50% (i.e. 4·5 min) to incorporate the larger volumes, because of slippage of the rubber/oil against the mixer rotors and chamber walls.

The procedures used and formulation evaluated have been mentioned above. As in that procedure the fatty-acid soap was added to the mixer in a small sachet after half the rubber had been added to ensure that intimate contact between the two was achieved with minimum transfer of the fatty-acid soap to the chamber walls.

4.3.2. Processability

As shown in Fig. 12 the reduction of viscosity by the use of fatty-acid soaps is significant and represents an effective means of obtaining a finished stock of required viscosity. The effect of 2 phr is quite dramatic, resulting in a drop of some 40–45 Mooney points. Levels above 2 phr do give further reductions, as Struktol A60 and Dispergum T best illustrate by their ability to give viscosities in the region of 30 Mooney points at a level of 10 phr. It is noted that increased dosages of Aktiplast or zinc stearate have little further effect on viscosity reduction. There may of course be problems with fatty acid soap bloom at higher dosage levels, although no such blooms were found during the laboratory trials. The lower viscosity level reached, when using these materials, is not accompanied by the stickiness caused by chemical peptizers. Fatty-acid viscosity reducers provide considerable reduction in processing time and greater machine utilization. Addition of oil to decrease the viscosity of the polymer is not an ideal method; it not only increases the time of breakdown, owing to slippage in the mixing chamber, but it also reduces the hardness of the final compound and adversely affects other vulcanizate properties.

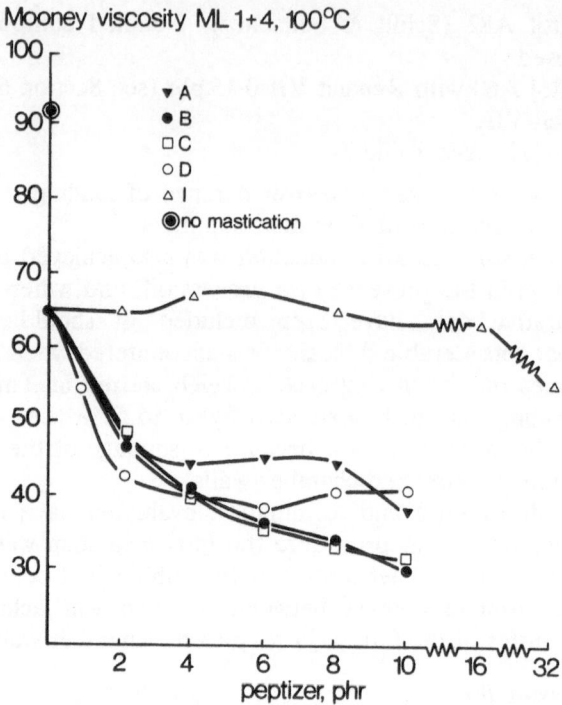

FIG. 12. Effect of peptizer level on Mooney viscosity. Conditions: K2A Intermix, 50 rpm, 100 psi, 3 min mastication.

4.3.3. Characteristics

Fatty-acid soaps have some inherent characteristics which make them more acceptable as a means of reducing compound viscosity than do chemical peptizers. Because of their fatty-acid soap base they could eliminate or reduce the need for added fatty acid activators, and considerably reduce the stickiness of low-viscosity natural rubber masterbatches. They can be used in a number of applications where conventional chemical peptizers could cause contamination problems, e.g. in the food industry. They must, however, be used in considerably higher dosages than chemical peptizers.

4.3.3.1. Internal lubrication. Unlike chemical peptizers these materials do not catalyse the breakdown of the backbone of the polymer chain during mechanical shearing, but appear to act as internal lubricants and thus enable a material of low viscosity to be produced without excessive breakdown.

This characteristic can be illustrated in the laboratory by dissolving natural rubber in a suitable solvent and blending the solution with solutions of the fatty-acid soaps of various concentrations, drying at the solvent boiling temperature and under vacuum to constant weight, and then determining Mooney viscosity and molecular weight of the polymer. The results of such an experiment are shown in Fig. 13. As can readily be seen, the viscosity of the polymer mass steadily reduces with fatty-acid soap addition whereas the molecular weight of the polymer, although subject to some scatter, remains at the same level as the untreated polymer.

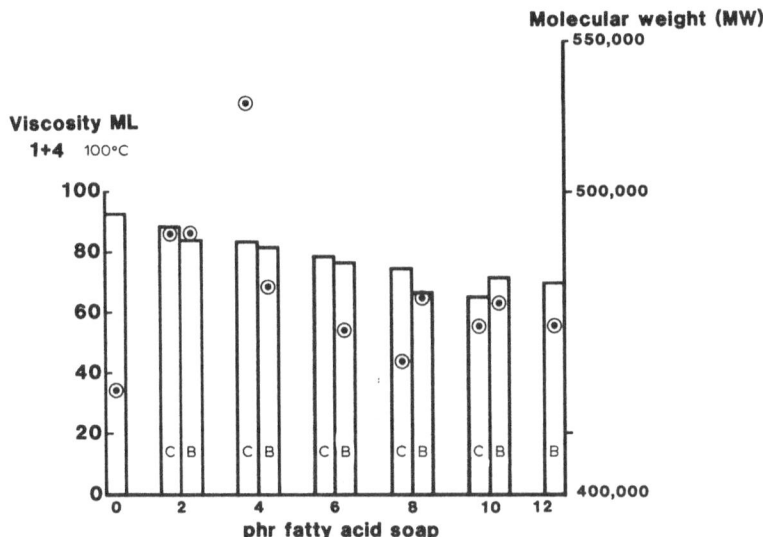

FIG. 13. Effect on viscosity of rubber induced by fatty-acid soap. ☉, Molecular weight.

4.3.3.2. Enhancement of green strength. The presence of a fatty-acid soap enhances the green strength of raw natural rubber presumably because of the phenomenon just described. At a given Mooney viscosity level the rubber containing the fatty-acid soaps can have a green strength 10% higher than that of masticated or chemically peptized rubber. This improvement is probably achieved because of less chainlength reduction.

4.3.3.3. Improvement of reversion resistance. A further interesting facet of fatty-acid soaps is their effect upon the reversion resistance of natural rubber compounds. If curemeter curves are compared for compounds containing conventional levels of stearic acid and those having additional fatty-acid soap it can be seen that as the level of fatty-acid soap increases the reversion of the compound on extended cure decreases, and can in fact be seen to effect a tightening of modulus when 5 phr of the fatty-acid soap are used (Fig. 14). This property may also be used to improve the ageing characteristics of the vulcanizate in circumstances when reversion of the polymer network affects the function of a component during its working life, e.g. in tyre tread formulations.

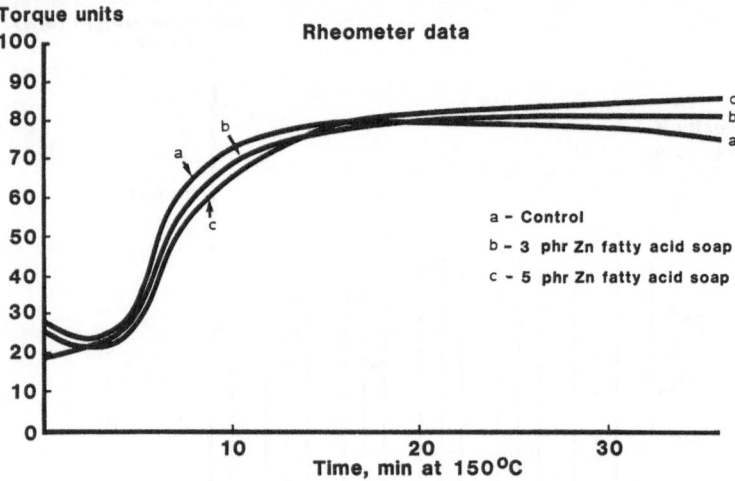

FIG. 14. Effects of fatty-acid soap on cure characteristics.

4.4. Blends of Fatty-Acid Soaps with Chemical Peptizers

These materials were introduced to the market in order to achieve the type of breakdown desired but not achieved by chemical peptizers or fatty-acid soaps alone. The blend should ensure rapid, even distribution of the chemical peptizer, eliminating the localized peptization often encountered with poorly dispersed material and resulting in a more homogeneous and less tacky peptized material. Better dispersion arises because of higher bulk of the blend, and lubrication of surface of the peptizer by the fatty-acid soap facilitates dispersion.

A further factor in the use of these peptizer blends is that a greater degree of batch-to-batch reproducibility should be possible, because the final viscosity is less dependent on the accurate weighing of very small amounts of peptizer than when chemical agents alone are used.

The methods of evaluation and compounds used for vulcanizate assessment were the same as those used above.

4.4.1. Viscosity Results

To put the acquired information into perspective it is necessary to consider the effects of simple mastication under similar conditions. The mastication of SMR 10 at high speed (50 rpm) in the K2A reduces the Mooney viscosity as shown in Table 9.

Using 1–2 phr peptizer blend the viscosity can be reduced in 2 min to a level similar to that achieved in 4 min mastication (see Fig. 15).

4.4.2. Mixing Energy

Table 9 shows distinct advantages for a fatty-acid soap/peptizer mixture, in this case an 'in-house' blend of 1–2 phr Struktol A60 and 0·15 phr Renacit VII. Although it shows some advantages in viscosity reduction and energy savings, this blending does have a distinct drawback in the factory, that of precision weighing of very small quantities of powders. It is vital to control this factor if wide variations in viscosity are not to occur. Any apparent cost savings from the use of in-house blends can soon be lost in this way and it is recommended that if a suitable commercial blend is available then, in terms of quality control, it is more economical to use this material even though it may be more expensive.

4.4.3. Vulcanizate Properties

The effects of this type of process aid on unaged and aged tensile strength are not very marked (see Fig. 16). The unaged data are very similar to data obtained from masticated rubber of similar viscosity. Similarly there is little effect upon elongation at break (Fig. 17). From the data shown, the fatty-acid soap/chemical peptizer blend appears to give a rapid and consistent reduction of viscosity in the processing of natural rubber. Ease of handling and reduction in power consumption, coupled with shorter processing cycles, give worthwhile savings. The use of these materials should be investigated by factories where considerably lower viscosities than those obtained by conventional mastication of raw natural rubber are required for subsequent ease of

TABLE 9
ENERGY REQUIREMENTS FOR VISCOSITY REDUCTION

Material	Dosage (phr)	Energy requirement (mJ/m^3)	Mooney viscosity, $ML1 + 4,\ 100°C$
Aktiplast	2	1051	48
	6	958	37
	10	840	37·5
Dispergum T	2	983	47
	6	856	36
	10	842	30
Struktol A60	2	985	49
	6	933	35
	10	814	32
Zinc stearate	2	941	43
	6	911	38·5
	10	864	41·5
Process oil	2	977	63
	8	822	56
	32	563	37
Dispergum 24	1·0	1029	56
	2·0	866	45·5
Struktol A82	0·8	969	42
	2·0	822	37·5
Struktol A60 +	1·0	891	54
(Renacit VII 0·1 phr)	2·0	861	45
Struktol A60 +	1·0	858	52
(Renacit VII 0·15 phr)	2·0	831	37

Mastication only	Time (min)	Energy requirement (mJ/m^3)	Mooney viscosity, $ML1 + 4,\ 100°C$
	2	722	74·5
	4	1421	67
	6	1905	43·5
	8	2266	29·5
	10	2528	21·5

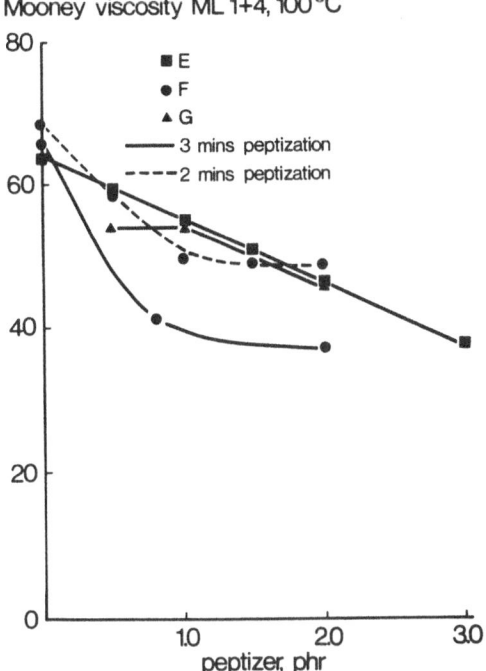

Mooney viscosity ML 1+4, 100 °C

FIG. 15. Effect of fatty-acid soap/chemical peptizer blends on viscosity. Conditions: K2A Intermix, 50 rpm, 100 psi, start 100°C.

processing. These materials have little apparent effect on physical properties.

5. PROCESS AIDS OTHER THAN VISCOSITY REDUCERS

Other types of process aids available to the rubber compounder to assist in various processing problems essentially fall into the following main categories.

(1) Esters of fatty acid soaps.
(2) Homogenizing resins.
(3) Factice.
(4) Superior processing natural rubbers.

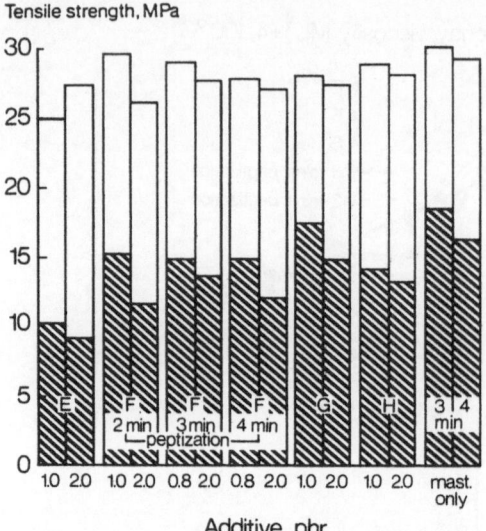

FIG. 16. Effect on vulcanizate tensile strength of peptization time, and type and dosage of peptizer. E, Dispergum 24; F, Struktol A82; G, Struktol A60/Renacit VII 0·1 phr; H, Struktol A60/Renacit VII 0·15 phr; □, unaged; ▨, aged three days at 100°C.

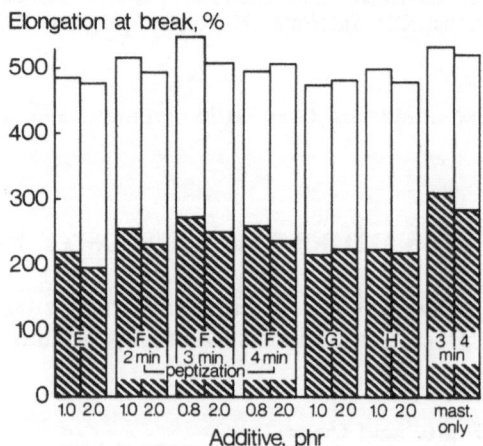

FIG. 17. Effect on vulcanizate elongation at break of peptization time, and type and dosage of peptizer. E, F, G, H, □ and ▨ as in Fig. 16.

5.1. Esters of Fatty-Acid Soaps

This class of material is utilized by the rubber compounder when he has a problem concerned with improvement of material flow in such products as intricate mouldings manufactured by compression or more often by injection moulding. To be economic, injection moulding needs compounds which will fill the mould rapidly and which will have consistent flow characteristics to ensure that all cavities of multi-cavity moulds are filled to make good products.

It is also claimed that fatty-acid esters enhance filler dispersion in compounds, thus achieving better physical properties from mixes. They also enhance extrudability of compounds, presumably by lubricating the rubber as it is sheared through an extruder. This feature is illustrated in Table 10.

Note that the low shear rate Mooney viscosity test does not show any advantage, and indeed in this instance suggests the reverse; yet in a high-shear high-speed extrusion test through a 2 mm diameter die obvious advantages are apparent for the process aid.

The esters can also assist in processing of very tacky compounds

TABLE 10

IMPROVEMENT OF EXTRUSION USING FATTY ACID ESTER

	A	B
Formulation (phr)		
NR, RSS1	100	100
Stearic acid	2	2
Zinc oxide	5	5
Antioxidant	1·5	1·5
N550, FEF black	35	35
N990, MT black	25	25
Sulphur	2·5	2·5
MBTS	1	1
TMTD	0·1	0·1
Fatty-acid ester (Struktol WB212)	—	2·5
Properties		
Mooney viscosity, ML1 + 4, 100°C	64	71
Extrusion rate (% improvement)	—	+12
Tensile strength (MPa)	19	19
Elongation at break (%)	470	430
300% modulus (MPa)	12	12
Hardness (° Shore A)	68	69

which stick to calender bowls, etc., but conversely they do not adversely affect adhesion of compounds to wire in tyre or cable applications.

5.2. Homogenizing Resins

These materials are relative newcomers and find their main application in the blending of dissimilar polymers and because of their nature they also assist in improving the dispersion of fillers as well as conferring tack to the polymer blend.

Homogenizing resins are low molecular weight polymeric resin blends and are of two types:

(a)   dark-coloured highly aromatic-based materials, which cause slight staining;
(b)   less polar, amber-coloured blends, which are non-staining.

Table 11 shows that a plasticizer is most effective when its solubility or polarity matches that of the rubber in which it is being used. This is taken to be a true solvating effect and not a pseudo-plasticization obtained by diluting the polymer with a large volume of incompatible liquid. The further apart any two types of rubber are in solubility parameter or viscosity, the more difficult it is to achieve compatibility. Blends of plasticizers, each compatible with different rubbers being

TABLE 11

COMPARISONS   OF   ELASTOMERS   AND   PLASTICIZERS: SOLUBILITY PARAMETERS

	Elastomer	Plasticizer	Homogenizing agent
	Urethane	Polar ethers	
	Nitrile (high-ACN)		
	Nitrile (medium-ACN)	Highly polar esters	
	Nitrile (low-ACN)		
Increasing solubility parameter	Polychloroprene	Low-polar ester	MS homogenizer
	SBR	Aromatic oils	
	NR		
	Polybutadiene	Naphthenic oils	
	Butyl		NS homogenizer
	EPDM	Paraffinic oils	
	EPM		

blended, are effective at improving blend homogeneity. However, plasticizer blends for homogenizing have the disadvantage of being prone to migration and bloom because one of the constituents will always be incompatible with one of the rubbers. To overcome this problem commercially available homogenizing agents contain higher molecular weight homologues of the plasticizers, i.e. resins, rather than liquids.

It has been claimed that they have a wetting effect on compounding ingredients and thereby reduce the energy necessary to get dissimilar components to mix more easily.

5.3. Factice
The use of factice as a process aid for rubber compounds dates back to the early days of the industry, when it was discovered that various vegetable and fish oils could be crosslinked by sulphur just as rubber could. The history of the factice industry is described as part of the proceedings of a symposium entitled *Factice as an Aid to Productivity in the Rubber Industry*[2] and the subject is historically discussed by R. F. Reynolds.[2]

5.3.1. Grades Available
Two main grades of factice exist, namely 'brown', which is usually vulcanized by elemental sulphur, and 'white', which is usually vulcanized by sulphur chloride.

Some suppliers still offer very extensive ranges of factices for specialized purposes. One supplier's range (DOG, Hamburg) consists of the following grades.

(1) White sulphur factice, chlorine-free.
(2) Yellow sulphur factice.
(3) Brown sulphur factice.
(4) White sulphur chloride factice for erasers and cold-cured rubber goods.
(5) White, sulphur chloride factices for heat-cured rubber goods, treated to remove labile chlorine.
(6) Factices for speciality elastomers.
(7) Sulphur-free and chlorine-free speciality factices, that are crosslinked by isocyanates or peroxides. They have excellent thermal stability compared with conventional factices.
(8) Low molecular weight factices which act during processing as

plasticizers but continue to crosslink during cure of the rubber. These find application in microporous rubber goods.

White factices, produced by normal methods of manufacture, are not entirely thermally stable and evolve HCl at vulcanizing temperatures. Similarly, if stored for any great length of time prior to incorporation in a rubber compound, HCl can be generated which will affect cure characteristics. If 'old' white factice is to be used, it should be steamed first to remove the HCl.

5.3.2. Factices as a Process Aid
There are two main areas in which it is claimed factice has a beneficial effect on rubber compounds: (1) during mixing, and (2) in subsequent processing and vulcanization.

If the factice is added in the early stages of a mix with the polymer, incorporation of the fillers is facilitated and certainly in the days when much of rubber compounding took place on open mills, the ability of the factice to ease incorporation of fillers and prevent 'bagging' of the mix (loss of adhesion of the mix to the mixing rolls) was a bonus point for its use. Factice also helps to prevent heavily filled stocks being internally mixed from crumbling and not forming a coherent mass.

The best known characteristic of factice, before the invention of partially crosslinked Superior Processing grades of NR and their synthetic equivalents, was the property of conferring dimensional stability to extruded and calendered stocks. Extruded items such as tubes, hose, cables and automotive window seals must retain their configuration during vulcanization either in open steam or during vulcanization by low-pressure processes such as LCM. In this area factice acts as a non-thermoplastic additive.

White factice did, and probably still does, find considerable use in the proofing industry. Rubber compounds containing white factice dissolve more readily to form doughs and produce more homogeneous and smoother materials. They also give more body to the spreading dough and show a marked tendency to reduce 'strike-through' and thus facilitate the attainment of a specified coating thickness with fewer applied coats.

5.3.3. Effects of Factice on Vulcanized Properties
Factices do have some small 'deleterious' effects on vulcanized physical properties. These may not be evident because factice can also

assist as a true process aid and enhance dispersion of ingredients, thus lifting properties like tensile strength. Tear strength of elastomers such as natural rubber is adversely affected but in the case of butyl rubber factice addition leads to substantial increases in this property. Similarly factice gives increased compression set with natural rubber but in loadings up to 15 phr in an EPDM compound can give considerable improvements, if certain special grades are used (DOG grades L900 and WP). Addition of factice to a compound allows one to reduce the level of the acceleration system somewhat, the factice contributing to the cure system in a way which has become known as cure compensation; properties such as tear strength, especially hot tear strength, are improved by such addition, e.g. 5 phr. Reductions of up to 40% of the accelerator are claimed possible by this method.

5.4. Superior Processing Natural Rubber

Superior Processing natural rubber is a polymer in which a proportion of the latex used in its preparation has been vulcanized in the latex phase prior to coagulation of the rubber to form the sheet or bale.

The inclusion of a prevulcanized phase in the polymer gives unique properties in that flow and collapse of an extrudate are greatly reduced; also the surface finish of products using SP rubbers is much smoother than those from conventional rubbers.

Various grades of SP are available with the proportion of crosslinked phase varying between 20 and 80%. Some may be used as a direct replacement for the usual polymer, whilst others are effectively masterbatches of the crosslinked rubber and can be used in combination with NR as well as with many types of synthetic rubbers.

SP rubbers find particular advantage when used in lightly loaded extruded or calendered products to the extent that even pure gum formulations can be processed as easily as heavily filled formulations. Particular benefits in extrusion are reduced die swell and faster throughput (Table 12). The effect of SP rubbers on vulcanizate properties is negligible.

Another useful feature is in soft stocks which have to be reworked without forfeiting the stiffness necessary for processing.

The proportion of vulcanized phase which should be used in a formulation is governed by the processing properties required. Often a proportion of 20 parts of vulcanized phase to 80 parts of unvulcanized rubber is adequate. This can be achieved either by complete replacement of the unvulcanized rubber component in the mix with an SP20

TABLE 12

COMPARATIVE VISCOSITIES AND EXTRUSION BEHAVIOUR[a] OF SP AND NON-SP RUBBER MIXES WITH DIFFERENT VOLUME LOADINGS OF CALCIUM CARBONATE

Volume filler (per hundred volumes rubber)	Mooney viscosity ML1 + 4, 100°C		Throughput[b] (g/min)		Die swell[c] (%)	
	RSS	SP-RSS	RSS	SP-RSS	RSS	SP-RSS
None	30	35	200	1000	49	25
20	38	54	320	940	31	20
50	40	55	420	600	20	11
80	75	90	560	530	14	6

[a] Data obtained with a 76 mm, variable-speed, hot-feed extruder, tubing die 29 mm external, 22 mm internal diameter.
[b] Screw speed adjusted to give the highest throughput consistent with good surface finish.
[c] Die swell measured as percentage increase of external diameter of tubing over that of die.

grade or by partial replacement with an SP40, SP50, PA57 or PA80 grade.

Where SP is to be used in soft oil-extended mixes, PA57 would be the natural choice as this provides a convenient method of adding oil to a mix.

Sulphur and accelerator levels should be adjusted to allow for the reduced amount of unvulcanized rubber in the formulation. With some vulcanizing systems mixes containing SP rubber may still cure faster and have shorter scorch times than normal mixes (see ref. 3).

6. PLASTICIZERS AND PROCESS AIDS—DO THEY REPRESENT USEFUL COMPOUNDING MATERIALS?

This chapter has dealt with a number of classes of the materials known to the rubber processing industry as process aids. The information drawn together emphasizes the importance of this group of materials. They represent very worthwhile and technologically useful solutions to a number of processing problems and should be included as a significant part of any compounder's armoury as the possible answer to a number of troublesome processing problems which occur daily in the running of processing factories. If process aids are used, many problems will be eradicated or at least significantly reduced.

ACKNOWLEDGEMENTS

The author wishes to thank the Board of the Malaysian Rubber Producers' Research Association for permission to publish this chapter, the editor of *Natural Rubber Technology,* Deutsche Oelfabrik GmbH and Schill & Seilacher for information and samples, and Mr P. M. Lewis for constructive criticism.

REFERENCES

Literature Cited

1. MORRIS, G., in *Developments in Rubber Technology*—1, ed. A. Whelan and K. S. Lee, Applied Science Publishers, London, 1979, Chapter 6.
2. *Factice as an Aid to Productivity in the Rubber Industry, Proc. Nat. Coll. Rubber Technol.,* London, 1962, p. 5.
3. *Natural Rubber Technical Information Sheet D47,* MRPRA, Hertford, UK, 1979.
4. *BP Process Oils,* publication No. SP/36/80, BP Oil Ltd, UK.

General References

CROWTHER, B. G., Effects of process oils on the vulcanizate properties of a natural rubber tread stock, *Nat. Rubber Technol.,* **14**(2), 1983, 79.
CROWTHER, B. G., Peptization of natural rubber in an internal mixer, *Nat. Rubber Technol.,* **12**(2), 1981, 27.
CROWTHER, B. G., Peptization of natural rubber in an internal mixer, Pt II. Fatty acid soaps and blends with chemical peptizers, *Nat. Rubber Technol.,* **14**(1), 1983, 1.
Factice DOG Kontakt 27, Deutsche Oelfabrik GmbH, trade literature.
Use of Homogenizing Resins (40MS/60NS) to Save Energy and Reduce Mixing Cycle, Bulletin No. 2, Schill & Seilacher, Hamburg.
Struktol Products for Tire Application, Schill and Seilacher, UK.

Chapter 5

A REVIEW OF ELASTOMERS USED FOR OILFIELD SEALING ENVIRONMENTS

W. N. K. Revolta and G. C. Sweet

Du Pont (UK) Ltd, Hemel Hempstead, Herts, UK

1. INTRODUCTION

The objective of this chapter is to give a general overview of elastomers used for sealing applications during the exploration and production of oil and gas.

The authors will also try to define the main problem areas and give broad distinctions between different elastomers in order to highlight their strengths and weaknesses.

2. OIL AND GAS PRODUCTION

The next few pages are designed to orientate those readers who have only a scant knowledge of the subject, and terminology, of oil and gas production. Readers wishing to expand their knowledge of oil and gas production methods further are recommended to consult refs 17 and 18.

2.1. Main Areas

The oil and gas industries (the Oil Patch) fall into two main areas, in so far as the use of elastomeric seals is concerned.

The first phase concerns exploration drilling, in which bore holes, often many hundreds of metres deep, are drilled into the earth in order to decide whether a suspected oil or gas 'reservoir' is worth further 'exploitation'. This phase has to be carried out, and is independent of

159

whether the well is to be sited on- or off-shore. The term 'drilling rig' is used to describe the equipment at this stage. Offshore, one refers to a drilling rig during the initial phase. Drilling offshore is carried out on various structures referred to as 'jack up rigs', 'semi submersible rigs' or simply 'drill ships'.

The next phase concerns production from the well. The term 'production rig' is then used to describe the structure and some of the components.

Between phases 1 and 2, some so-called developmental drilling is carried out on those discoveries which are highly likely to yield commercial quantities of oil or gas.

The term 'platform' is only used to describe a development and subsequent production facility, offshore. It is a much more massive structure than that used simply for drilling and can be fixed to the sea bed or floating.

From the point of view of elastomer seal usage, Table 1 (see later) lists the types of seal which apply to drilling and/or production. However, there is an overlap of use in each phase. Further details will follow later in this section.

2.2. Drilling

Drilling is carried out by means of a rotating drill which is attached to lengths of drill pipe. Lengths of pipe are continually added as drilling proceeds.

Drill mud is pumped down the inside of the pipe by a circulation system in order to allow removal of the drilled rock fragments. Additionally the mud, which is composed of oil, water and heavy minerals (e.g. barium sulphate), provides a pressure, to prevent reservoir fluids blowing out of the hole. Other chemicals which give problems for the various seals already used at this stage, e.g. corrosion inhibitors, viscosity modifiers, etc., are added to the mud.

Even before the well starts to produce, a very important valve system is placed at the top of the hole. This is the 'blow out preventer' stack. The blow out preventer, as the name implies, prevents the oil or gas blowing out of the well inadvertently due to a 'kick'. This happens if some geological effect overcomes the pressure afforded by the weight of the drilling mud.

Blow out preventers (BOPs), either annular or ram-like in operation, contain very large rubber sealing elements. When activated these elements can close extremely rapidly to seal off a hole.

A choke manifold, which consists of a series of remotely controlled valves, allows the driller to control and relieve downhole pressures, following a 'kick'.

2.3. The Well and its Components

As the well is drilled, lengths of large-diameter steel pipes are let into the hole and cemented in. This is called the 'casing' or 'casing string' and serves to reinforce the bore hole. If it is decided that a well will actually go forward as a production prospect, this casing goes to the bottom of the hole and below the area of rock in which the reservoir is to be found. In order to obtain the oil, gas or oil–gas mixture (condensate) holes are literally blown through the casing at its bottom end, by explosive charges.

Production tubing is then let into the casing string. This is the smaller-diameter steel tubing that will actually carry the reservoir fluid and gas up to the surface.

Between this tubing and the casing is placed a 'packer'. The packer and its sealing element forces reservoir fluid to run up the tubing rather than up the annulus between the tubing and casing. It is not desirable to produce from the well via the casing.

Packers are either permanent, remaining in position throughout the life of the well, or retrievable. This difference depends upon whether the tubing needs to be withdrawn or not. As will be seen later, the composition of the rubber sealing element of packers differs, according to ultimate use.

Tubing is hung on a bracket support system called a 'hanger'. 'Hanger seals' are very large, square-cut pieces of rubber used to prevent fluid or gas escaping at the point of attachment.

When the production tubing string has been run, and the packer 'set', i.e. it is allowed to seal against the casing, the well is almost complete. The final task is to put a 'Christmas tree stack' at the top of the bore hole. This is a series of control valves used to control and deliver the products of the well. No-one knows now why this name is used, unless it happened to look, at one time, like a Christmas tree bedecked with coloured lights. Many seals are used in this system.

In offshore production, this unit is placed just below the platform and is fed by the 'marine riser', i.e. a large flexible pipe which carries oil or gas from the sea bed up to the surface.

Obviously, a large number of other non specific 'O'-rings, chevrons and square-cut seals find their way into this complex system of oil exploration and production equipment.

2.4. Well Enhancement or Stimulation

Oil is found within the pores of rocks, and only a proportion of this will flow under natural pressure. In order to obtain the maximum amount of recoverable oil from a well, it is often necessary to 'stimulate' it to give up more of what it contains.

Several means have been developed to achieve this end. They include:

(1) Acid stimulation, which dissolves away certain rock formations.
(2) Hydraulic fracturing, i.e. breaking the rocks by pumping a special fluid mixture at very high pressure.
(3) Gas injection: normally local natural gas, although nitrogen–carbon dioxide is also used.
(4) Water injection.
(5) Steam injection.

2.5. Artificial Lift

This is the process used to lift oil by other methods when natural pressure is insufficient. High-pressure gas and surface or submersible pumps are employed.

2.6. Pipe Lines

In the production of oil and particularly of gas, long pipe lines have to remain clean on their inside bore. In order to achieve this, devices known as 'pigs' are sent down the lines by gas pressure. Pigs are essentially circular scrapers. So-called 'pig release valves' send the pigs on their way and the seals have to be unaffected by rapid changes in pressure at this point of release.

3. SEAL APPLICATIONS AND ENVIRONMENT

3.1. Seal Applications

The general range of application areas of elastomers and some plastics in seals is given in Table 1.

3.2. Environment

In the applications specified in Section 3.1 the seals, in whatever form, are required to function in a wide range of fluids, temperatures and

TABLE 1

APPLICATIONS OF ELASTOMERS AND PLASTICS IN SEALS

Drilling	Production
Pump valves and pistons	Packers Permanent Retrievable
Valve liners	Blow out preventers (BOPs)
Drill head seals	Safety and choke valves Casings and hangers Christmas trees Pig release valves

pressures, and in a variety of media: short lists are given in Tables 2 and 3.

The capability of elastomeric sealing materials is often stretched to the limit under such environmental combinations, and the engineer will often prefer to use a metallic seal. However, elastomeric seals are

TABLE 2

POTENTIAL OILFIELD ENVIRONMENTS: PHYSICAL CONDITIONS AND NATURAL MEDIA

Temperatures	200°C and above Arctic capability
Pressures	>15 000 psi (1020 bar)
Natural media	Crude hydrocarbons $\left.\right\}$ or condensate Hydrocarbon gases H_2S, CO_2, water, brine

TABLE 3

POTENTIAL OILFIELD ENVIRONMENTS: DRILLING, PRODUCTION AND ENHANCEMENT MEDIA

Drilling media	Muds—oil/water based Completion fluids—heavy salt Treatment fluids—strong acids
Production media	Corrosion inhibitors—amines/high pH
Enhancement media	Steam Liquid nitrogen Carbon dioxide Water

still an essential component of most systems and this is likely to be the case for the foreseeable future.[1]

4. ELASTOMER SELECTION

As an aid to making a preliminary selection, the well-known rubber industry ASTM Classification System has been used. Originally this was designed for the automotive industry in the USA.

4.1. The ASTM Classification

The oil and heat resistance of some polymeric materials is shown in Fig. 1.[2]

This approach to selection is of interest for two main reasons. Firstly, it does rapidly define those elastomers which have oil resistance, a prerequisite for the oil industry. It then defines the maximum temperature rating for service conditions. For example, the HK category relates to fluoroelastomers, an FE rating to silicone rubbers, and so on. Refer to ASTM D2000[2] for a full list of codings. Secondly this chart is an example of how elastomers might eventually be classified for the oil industry. It would be very desirable for both the seal supplier and the user, if a simple system such as this could be introduced to at least narrow the field. For example, if resistance to sour oil was used as one parameter (instead of swell in ASTM3 oil which is an automotive test oil), the chart would start to have some relevance to the oil industry. This theme will be expanded later (Section 6).

As can be easily seen, the materials to the right of the chart, i.e. those which have best oil resistance, are candidate elastomers for oilfield service. Then, moving to the top right of the chart, we see which of these materials combine oil resistance with stability at the temperature envisaged.

Plastics such as Teflon®, PTFE and FEP may be included here, since they find use as harder, back-up rings, in combined sealing elements in the harshest environments, and where gas decompression problems may occur.

Those materials which fit generally into the area bounded by 60% maximum oil swell and 100°C minimum temperature resistance will now be reviewed.

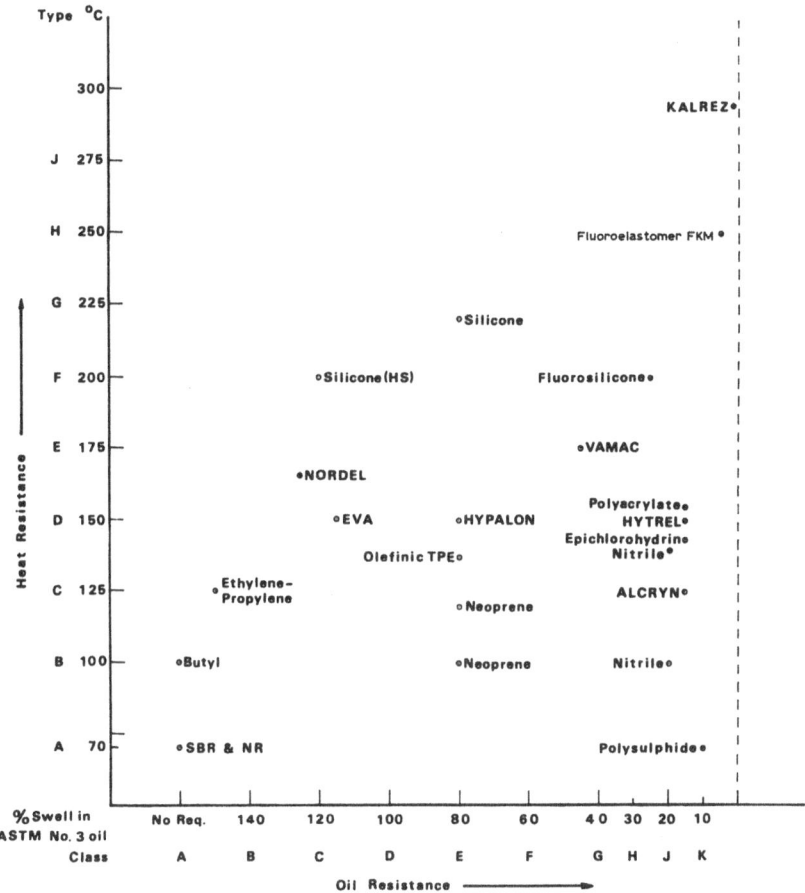

FIG. 1. Oil and heat resistance of some seal materials.

Notes

(1) The SAE J200/ASTM D2000 classification only covers crosslinked elastomers. Hytrel, Alcryn and the other thermoplastics are included for comparison.

(2) The temperature of testing in ASTM No. 3 oil varies with the heat resistance classification (type) as follows:

Type	A	B	C	D–J
Test temperature (°C)	70	100	125	150

(3) Specific points are shown for clarity. In practice, there will be a 'spread' depending upon compounding and grades of polymers used.

4.2. Comments on Specific Materials

4.2.1. Nitrile Rubber (NBR)

The 'workhorse' of the industry, NBR is considered very adequate as a seal material up to and around 100°C. It falls down badly at temperatures in excess of this, and in particular in sour oil environments.[1] In spite of this, permanent packer elements of nitrile work very well. In fact the hardening that occurs in service at higher temperatures actually seems to benefit the well system.[1] For retrievable packers, where flexibility must be maintained, other elastomers are now being increasingly used. Nitrile rubber is also used regularly in BOP seal systems, both ram and annular types.

In the opinion of one of the major manufacturers of BOP stacks, these very large seals normally fail as a result of mechanical damage. Thus the ability of special elastomers to resist more aggressive fluid environments is less important than high mechanical strength. The shear bulk of these seals limits chemical attack to the surface only.

It is interesting to note that, until a few years ago, natural rubber was used for BOP seals.

4.2.2. Carboxylated Nitrile Rubber

This material is also now being considered for oil field service, possibly because of its exceptional abrasion resistance and higher mechanical strength at elevated temperatures. The resistance of carboxylated nitrile rubber (XNBR) to hydrogen sulphide will increase the evaluation of XNBR for oil well usage. Overall heat and oil resistances, however, are similar to those of standard nitriles.[3]

4.2.3. Epichlorhydrin (ECO)

This appears to have some merit and the homopolymer version has exceptionally low gas permeability.[4]

4.2.4. Polyacrylate Rubber

Polyacrylate rubber, on the face of it, also looks a good candidate, but it is badly affected by hydrolysis in aqueous environments. This may account for the fact that it seems to be used hardly at all in the oil industry.[4]

4.2.5. Fluorosilicone Rubber

This rubber is not used, in spite of its good heat and oil resistant properties, because of low mechanical strength.

4.2.6. Fluorocarbon Rubbers

Examples of such materials are Viton® (Du Pont) and Fluorel® (3M). These appear to offer the best alternative to nitrile rubbers since they give a major improvement in resistance to very hot, sour conditions. This, combined with good compression set, adequate low-temperature properties and a broad range of fluid compatibility, apart from amine incompatibility (at least with standard forms), has led to their increasing use in oilfield environments.

4.2.7. TFE/Propylene Rubber[5]

This material (TFE/P) has good heat and oil resistance although inferior to fluorocarbon rubbers. It does, however, have good resistance to amine-based corrosion inhibitors and for this reason is finding some special applications. Wider use seems restricted due to relatively poor low-temperature properties and poorer compression recovery compared with fluorocarbons and nitrile.

4.2.8. Perfluoroelastomer (PFE)

Kalrez® (Du Pont) elastomer offers complete resistance to well fluids, at temperatures exceeding even those of fluorocarbons. Unfortunately, since this material is a partial thermoplastic, it suffers from poor resistance to extrusion and has to be backed up by harder materials, e.g. Teflon®, Ryton®, etc. Textile fibre filled grades of Kalrez have shown a great improvement in extrusion resistance.

Although this polymer is more expensive than any other candidate for really long-term, down-hole service life, Kalrez® PFE is unsurpassed.

4.3. Low-Temperature Service

There are of course some applications in the oil industry which require a seal to operate under so-called 'Arctic' conditions. Thus it is useful to look at the chart in a different way. In Fig. 2, oil resistance is plotted again, but this time against *minimum* service temperature.

Bearing in mind what has already been said about the heat and oil resistance of these materials, one can see why nitrile and fluorocarbon rubbers tend to be the mainstay of the industry.

Note that epichlorhydrin (ECO) has good low-temperature properties in the co-polymer (C) form, but not in the homopolymer (H) form. Unfortunately it is the H form which gives very low gas permeability.[4]

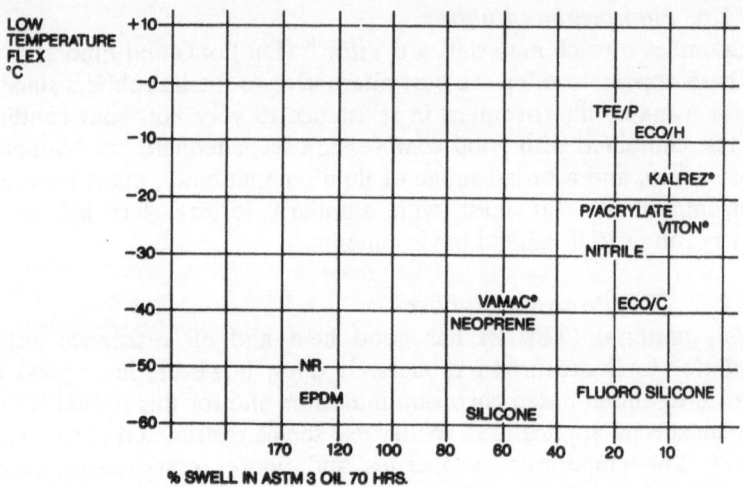

Fig. 2. Low-temperature service versus oil resistance of elastomers.

As already mentioned, TFE/propylene rubber is quite poor, even at around 0°C.

It is an unfortunate fact of polymer chemistry that elastomers with good oil resistance are generally poorer at low temperatures than their counterparts which have poor oil resistance.

5. SPECIFIC PROBLEM AREAS FOR ELASTOMERIC SEALS

Some particular seal compatability requirements have already been mentioned. Table 4 lists these for reference, not necessarily in any order of priority.

TABLE 4

SEAL PROBLEM AREAS

Explosive decompression
High-pressure extrusion
Effect of corrosion inhibitors
Effect of hydrogen sulphide
Steam/acid resistance
Methanol resistance

5.1. Explosive Decompression

The sudden decompression of an elastomeric seal from a high-pressure gas environment can lead in most elastomers to the production of blisters, internal and longitudinal surface splits. These features, produced under the above conditions, are described as being caused by explosive decompression.

The mechanics of the problem was investigated by Ender,[6] who noted the different swelling of elastomeric seals by different gases; carbon dioxide and hydrogen sulphide produced large swelling in several elastomers used in oil-filled environments. The use of methane led to considerably less swelling, showing that the type of gas in the high-pressure environment was important.

Further exhaustive work carried out by Potts[7] and Cox[8] examined the broad range of elastomeric candidates for oilfield use.

These reports of laboratory testing combined with field experience showed that fluoroelastomer outperformed the other leading candidates, viz. nitrile, carboxylated nitrile, hydrogenated nitrile and epichlorhydrin, especially when the overall oilfield environment was considered.

The further factor which was seen as a cause or source of explosive decompression was the presence of inhomogeneities or microporosity in the cured elastomer. High mechanical strength of the elastomer, its modulus and inherent viscosity are most important although the optimum value is seen to be below the maximum value obtainable from the elastomers.

5.1.1. Explosive Decompression Resistance

Correct choice of elastomer as the base for the seal compound is most important, as is the correct choice of compounding ingredients. The use of a reinforcing carbon black of optimum particle size in conjunction with high crosslink density maximizes the resistance to explosive decompression.[8]

The main factors to consider in choosing the compound material from which the seal is to be made are

(i)   choice of elastomer,
(ii)   high modulus/high crosslink density,
(iii) optimum compounding—especially filler choice,
(iv)   gas composition and temperature,
(v)   gas pressure and its fluctuations,

(vi)   cross-sectional area of seal and
(vii) high quality—elimination of inhomogeneity and micro-
        porosity.

Several commercial seal compounds based on FKM fluoroelastomer
have now gained acceptance in this exacting condition.[8,9] Field
experience has been gained with seals in such applications as process
ball valves and pig launcher units.

5.2. Resistance to Extrusion of Seals Under High Pressure

A seal should be able to resist extrusion out of its containment groove
when subjected to high pressures and temperatures. This property is
a combined function of the effect of toughness, stiffness at high temperature and hardness.

Design of the seal grooves is an important factor where reduced gap
dimensions minimize extrusion. The thermal volumetric expansion of
the elastomeric seal should also be allowed for, with special consideration to the expansion characteristics of the type of elastomer being
used.

5.2.1. The API Test

Evaluation of the extrusion resistance of an elastomer is facilitated by
a useful test method developed by the American Petroleum Institute
(API) in the 1950s.[10] The test is run on a standard ASTM compression
set pellet. One modification was the drilling of a hole in the tester
body for a thermocouple so that the test temperature could be
accurately controlled. The standard annular extrusion gap set up by
the API was 0·34 mm (0·0135 in) on each side, as shown in Fig. 3.

This test method was modified by Stevens of Du Pont to give a more
stringent but meaningful test.[11] Figure 3 shows the apparatus. The test
involves compressing a rubber pellet, so that it can flow past a
standard annular extrusion gap of 0·34 mm. Different pressures and
temperatures can be applied.

The procedure is carried out by placing the API extrusion tester in a
laboratory press, then heating the device until a temperature of 204°C
(400°F) is reached on the thermocouple in the body of the tester. Then
the rubber pellet, which has been previously weighed, is conditioned
in the tester for 5 min. A predetermined constant pressure varying
from 34·5 MPa (5000 psi) to 207 MPa (30 000 psi) is then applied to the

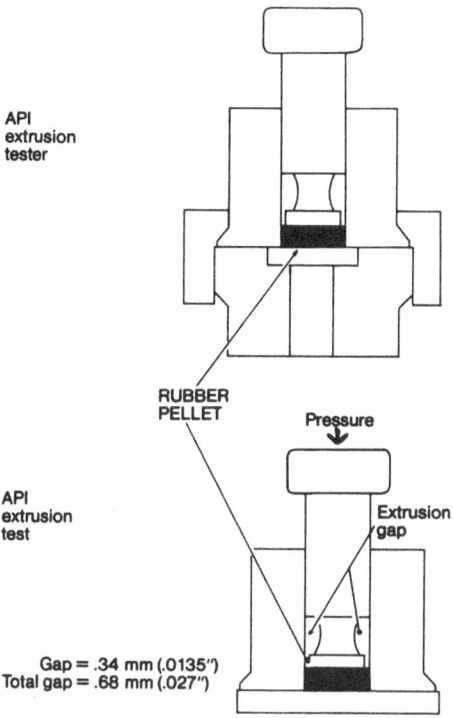

API
extrusion
tester

RUBBER
PELLET

Pressure

API
extrusion
test

Extrusion
gap

Gap = .34 mm (.0135″)
Total gap = .68 mm (.027″)

FIG. 3. API extrusion test apparatus.

piston. The deflection of the piston is measured for 5 min at one-minute intervals. The apparatus is then opened and the remaining pellet weighed after cooling. The total deflection is used to calculate the percentage retention of the original height of the pellet which is used as the measure of extrusion resistance as shown in Table 5.

This test method was designed to be quick so that numerous samples can be run, yet accurate enough to give meaningful data. An arbitrary 'rating scale' was then devised based on deflection test data so that a simple judgement could be made easily on the extrusion resistance of the rubber compound tested.

As pressure increases, at a given temperature, extrusion occurs and the pellet height decreases. The remaining pellet height is used as an arbitrary measure of extrusion resistance.

TABLE 5
EXTRUSION TEST CALCULATIONS

Extrusion resistance	Rating scale Original pellet height (%)[a]	Deflection (mm)
Excellent	87–100	≤1·65 (0·065 in)
Good	74–86·9	1·67–3·3
Fair	50–73·9	3·32–6·35
Poor	<50	>6·35 (0·250 in)

[a] Percentage of original pellet height $= \dfrac{(12·7 - d)}{12·7} \times 100$ where d = deflection.

5.2.2. Results of API Test

In the evaluation commercial 90 Shore A hardness extrusion resistant compounds based on NBR were compared with fluoroelastomers, normally or originally considered to have poor resistance to extrusion. The formulations of the compounds evaluated are tabulated in Table 6.

A plot was made of the percentage pellet height remaining compared with the pressure applied (see Fig. 4) examining two very good nitrile compounds compared with a number of fluoroelastomer compounds (see Table 6). At a temperature of 200°C the degree of extrusion (loss of pellet height) increases with pressure for all the samples tested.

After looking at this chart, one can speculate why confusion still exists about the performance of fluoroelastomers on extrusion at high pressures. If an end-user tested a compound based on the 1960s type of compounding such as is seen in formulation C, the conclusion might easily be that fluoroelastomers are much weaker than nitrile. Even a high-quality 90 Durometer O-ring stock such as D falls short of the nitrile performance. If, however, the nitrile is compared with an FKM AHV compound such as E, the fluoroelastomer stock compares favourably. These data demonstrate that fluoroelastomers can vary widely in performance and the proper compounding is necessary to ensure successful use.

One other point noted in this chart is the performance of peroxide-cured fluoroelastomers. While one of the best overall peroxide

TABLE 6

COMPOUNDS USED IN EXTRUSION RESISTANCE EVALUATION USING API TEST PROCEDURE

Polymer	A	B	C	D	E	F
NBR 1051	100	100	—	—	—	—
FKM A	—	—	100	—	—	—
FKM E60C	—	—	—	100	—	—
FKM AHV	—	—	—	—	100	—
FKM GF	—	—	—	—	—	100
FKM Curative Masterbatch No. 20	—	—	—	—	2	1·7
FKM Curative Masterbatch No. 30	—	—	—	—	6	4
Magnesia—low activity	—	—	15	—	—	—
Magnesia—high activity	—	—	—	3	3	—
Zinc oxide	1·5	1·5	—	—	—	—
Sublimed litharge	—	—	—	—	—	6
Calcium hydroxide	—	—	—	6	2	2
Process aid (NBB)	1·5	1·5	—	—	—	—
Stearic acid	0·5	0·5	—	—	—	—
FEF Black (N550)	60	60	—	—	—	—
SRF Black (N762)	60	60	—	—	40	40
MT Black (N990)	40	40	60	60	—	—
Dioctyl phthalate	5	5	—	—	—	—
Antioxidant TMQ	2	2	—	—	—	—
Sulphur	1·5	—	—	—	—	—
TMTD	2	—	—	—	—	—
TBBS	1	—	—	—	—	—
Scorch retarder	0·5	—	—	—	—	—
TAC	—	1·5	—	—	—	—
Diak No. 7 (TAIC)	—	—	—	—	—	3
Dicumyl peroxide, 40%	—	5	—	—	—	—
Varox 50%	—	—	—	—	—	3
Process aids (Viton®)	—	—	—	—	1	1·3
Diak No. 3[a]	—	—	3	—	—	—
Polyethylene 617A	—	—	1	—	—	—

[a] N,N-Dicinnamylidene-1,6-hexanediamine.

compounds, F, is superior to standard 1960s fluoroelastomer compound C, it falls short of nitrile's extrusion resistance or high-quality dipolymer/bisphenol cure technology (D and E). It should be remembered, however, that the peroxide-curable fluoroelastomers have superior steam and acid resistance to bisphenol cured FKM AHV, and

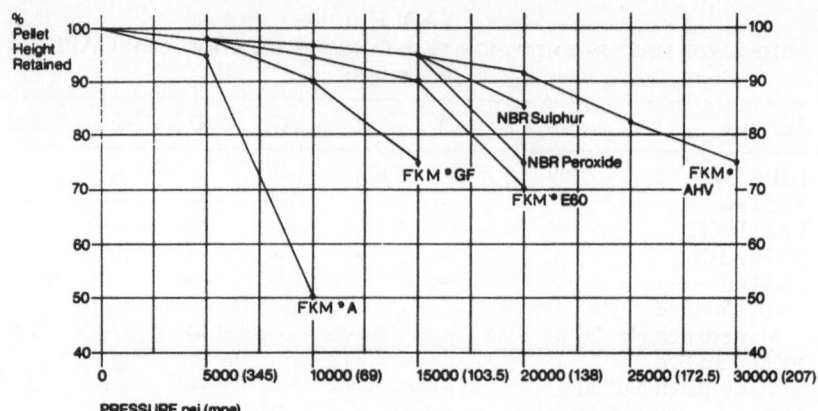

FIG. 4. API extrusion resistance tested at 200°C.

that they may be the polymers of choice if the application is under only moderately high pressure but chemically aggressive conditions.

The interesting conclusion was that although conventional 'automotive'-type fluoroelastomer compounds were certainly poorer than nitrile, the high-modulus FKM AHV compound performed best of all. Hence, providing proper selection is made, fluorocarbon rubbers can be considered to be at least as good as nitrile in extrusion resistance.

This test could be used as the basis of a screening test for extrusion resistance in a specification for elastomeric parts for oilfield performance (see Section 6).

5.3. Effect of Oil Well Additives
Chemical additives such as corrosion inhibitors and highly acidic or basic completion fluids can have serious effects on elastomeric components.

Unfortunately there is no universal elastomer that will handle all potential environments and the range of proprietary oil well additives is large. Furthermore, working practices in oil wells differ.

For example, cases have been reported[12] where identical seals failed in one well but performed perfectly in another. The difference, discovered after the event, was simply that corrosion inhibitor was used in one case and not in the other, even though both wells were completed by the same operator.

Thus a conventional nitrile seal may perform adequately in some situations but fail badly in others. Even fluoroelastomers can fail in some cases if not based upon the correct polymer type or curing system.

It does not make economic sense to use seals which are 'over-engineered' for the task in hand. However, an expensive workover can easily nullify the savings made on original equipment. Thus it is important for tool suppliers, elastomer seal moulders, and the well operator to understand what the options are.

5.3.1. National Association of Corrosion Engineers (NACE)
NACE recognized the problem of trying to evaluate elastomer performance in contact with corrosion inhibitors. Short of testing each elastomer, in each proprietary inhibitor, there was no way of generalizing. Hence NACE put forward two standard inhibitor formulations and called them NACE A and B.

NACE A is a water-soluble inhibitor and is composed as follows:

0·2 g-equivalent acetic acid,
0·1 g-equivalent N-COCO-1,3-propylenediamine,
20 g isopropyl alcohol,
30 g water.

NACE B is an oil-based inhibitor and has the following composition:

0·1 g-equivalent vegetable residue acid (dimer/trimer acid number 150),
0·1 g-equivalent of N-tallow-1,3-propylenediamine,
0·01 mol nonylphenol/9–10 mol ethylene oxide, condensation product,
100 g heavy aromatic solvent (90% distillable below 260°C).

These standard inhibitor solutions have been used over the last two or three years to evaluate the performance of some oilfield seal elastomers and some of these data will be reviewed here.

5.3.2. Comparison of NBR with Fluoroelastomers
A paper by Du Pont[13] published in 1984 compared nitrile elastomer with various fluoroelastomers. The results compared soft and hard seals at 150°C and 200°C, in various oil well fluids.

A rating system was used to assess performance after the various ageing periods, i.e.

Rating	Property retention (%)	Volume increase (%)
Very good	90–100	0–10
Good	70–90	10–20
Satisfactory	50–70	20–25
Fair	25–50	25–30
Poor	25	30
Brittle	No property retention	—

Results for the soft (70–80) nitrile and fluoroelastomer materials are shown in Table 7, for an ageing period of three days at 150°C. They are in some respects surprising. They show quite clearly that nitrile rubber is very poor in contact with NACE B and an acid completion fluid (results at 100°C were the same, thus eliminating the possible effect due to temperature). On the other hand, nitrile is actually better than fluoroelastomer in contact with NACE A. It also works quite well in highly basic completion fluid.

Although not recorded in this chapter, the original reference shows the same data for hard (90–95) seals. The outcome was essentially the same except that all the fluoroelastomer grades were somewhat better in all fluids.

What is quite apparent from these results is that FKM GF, a peroxide-cured, highly fluorinated type, has better overall resistance to all fluids than the Viton® A and B types.

Table 8 shows results obtained on Viton® seal compounds in the same fluids, at 200°C. Nitrile could not perform well at this temperature due to simple heat degradation.

TABLE 7

EFFECT OF OIL WELL ADDITIVES ON ELASTOMERS (THREE DAYS AT 150°C)

Test fluids	Elastomer type			
	Nitrile	FKM A	FKM B	FKM GF
Sour oil + 5% NACE B	Brittle	Poor	Poor	Satisfactory
Water/mud + 5% NACE A	Good	Fair	Fair	Satisfactory
Completion fluid, pH 11	Satisfactory	Fair	Fair	V. good
Completion fluid, pH 2	Brittle	V. good	V. good	V. good

TABLE 8

EFFECT OF OIL WELL ADDITIVES ON FLUOROELASTOMERS (THREE DAYS AT 200°C)

Test fluid	Elastomer type				
	FKM A	FKM B		FKM GF	
	Soft only	Soft	Hard	Soft	Hard
Sour oil + 5% NACE B	Poor	NA^a	NA	NA	Fair
Water/mud + 5% NACE A	Fair	Fair	NA	Satisfactory	Poor
Completion fluid, pH 11	Fair	Fair	Fair	Good	Satisfactory
Completion fluid, pH 2	VG^b	VG	VG	VG	NA

^{a} NA, no data.
^{b} VG, very good.

Unfortunately several data points are missing. However, one can conclude that, as before, fluoroelastomers have excellent resistance to acid completion fluid even at this extreme temperature. FKM GF again provides good resistance to basic completion fluid and also appears to give fair to satisfactory performance in both NACE A and B fluids. The hard version of this compound, for some reason, is poor with NACE A fluid.

Bearing in mind that at high temperatures and (as will be shown later) in very sour conditions, nitrile is essentially eliminated as an option, then the GF type FKM shows reasonable overall performance in the presence of additives.

5.3.3. Comparison with Other Systems

A useful paper by Watkins[12] has confirmed some of the data given in the previous section and provides information on other elastomers. Table 9 gives some additional data on the difference between a fluoroelastomer cured with peroxide and one cured by a conventional bisphenol system. The data confirm the Du Pont results, i.e. that a peroxide cure system gives better results.

Further data from the Watkins paper have been summarized in Fig. 5. This rates a number of elastomers according to cumulative change in physical properties as an index of total change (example: tensile strength − 20%, elongation − 30%, hardness + 10; total index 60 points change).

TABLE 9

EFFECT OF CURE SYSTEM ON FLUOROELASTOMER PERFORMANCE IN CORROSION INHIBITORS (SEVEN DAYS AT 100°C)

Property	Inhibitor			
	5% NACE A		5% NACE B	
	Cure system		Cure system	
	Bisphenol	Peroxide	Bisphenol	Peroxide
Hardness change (IRHD)	−13	−6	−5	−7
Tensile strength change (%)	−52	−34	−53	−44
Elongation at break change (%)	−29	−17	−35	−4

Of the different materials shown in Fig. 5 the most promising appears to be the TFE/propylene polymer Aflas. This is followed by hydrogenated nitrile (HNBR) and fluoroelastomer, which in this case was a conventional type. As we have seen, an improvement would be expected for a peroxide-cured version.

Conventional nitrile shows an overall property change index of 40 but carboxylated nitrile (XNBR) nearly 60—not a good recommendation.

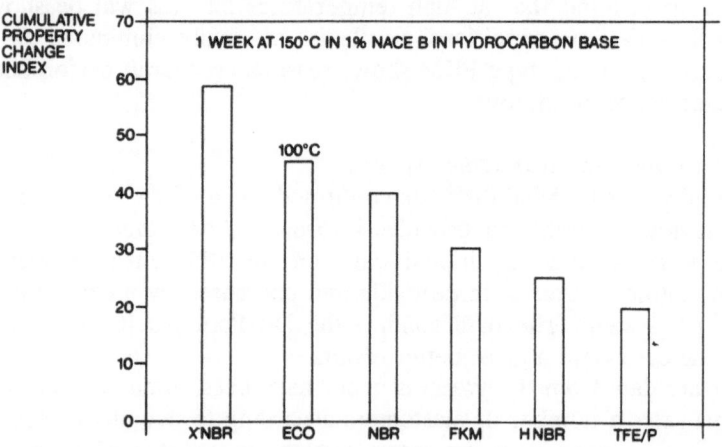

FIG. 5. Comparative effect of amine-based corrosion inhibitor on elastomers kept for one week at 150°C in 1% NACE B in a hydrocarbon base.

TABLE 10
TFE/P COMPOUND BASED ON AFLAS 100H

Property	Original	70 h/150°C in drilling mud + 5% NACE A	70 h/200°C in diesel mud + 5% NACE B
Hardness (IRHD)	95	92	78
Elongation at break (%)	90	140	130
Tensile strength (MPa)	21·5	18·5	14·7
Volume swell (%)	—	3·7	28·0

Note that the results for epichlorhydrin rubber (ECO) are only for 70 h at 100°C and one would therefore not expect this polymer to be as good as nitrile in a direct comparison.

5.3.4. Some Fluoropolymer Results

Further reference to TFE/P and corrosion inhibitor performance is given in a paper by Hull.[5] Results for Aflas type 100H in NACE A and B are given in Table 10.

TFE/P is also quoted[5] as performing well in a high-pH completion fluid at 175°C, showing similar performance to Viton.®

Perfluoroelastomer (e.g. Kalrez®) seals give excellent performance in amines since the material is almost entirely chemically inert. Table 11 shows the effect of highly concentrated inhibitor and the minimal change which occurs.

In spite of their high cost Kalrez® seals are always used where guaranteed long-term service in a combination of chemically severe environments, i.e. amines, sour gas, steam, very high temperatures, etc. is required. Such seals are often used in combination with PTFE

TABLE 11
RESISTANCE OF PERFLUOROELASTOMER TO AMINE-STABILIZED OIL

	Volume swell (%)	tensile strength retained (%)	Elongation retained (%)
20% Kontol®[a] IIIX in Canthus®[b] 210 oil (7 days, 140°C)	2	108	95

[a] Petrolite
[b] Exxon

TABLE 12

ELASTOMER COMPATIBILITY WITH OIL WELL ADDITIVES

Elastomer	Corrosion inhibitor				Completion fluid			
	Amine, oil base		Amine, water base		Acidic		Basic	
	150°C	200°C	150°C	200°C	150°C	200°C	150°C	200°C
Nitrile	F^a	F	G^b	F	F	F	G	G
Hydrog. nitrile	G	F	G	F	NA^c	NA	NA	NA
Standard FKM	F	F	G	G	G	G	G	F
Peroxide FKM	G	G	G	G	G	G	G	G
TFE/propylene	G	G	G	G	NA	NA	G	G
Perfluoro	G	G	G	G	G	G	G	G

^{a} F, Expected to fail.
^{b} G, Expected to give good service.
^{c} NA, Data not available.

or polyphenylene sulphide (Ryton) back up rings, or with FKM as the primary activator.

5.3.5. Recommendations

The foregoing data allow us to make some reasonable conclusions and recommendations concerning elastomer compatibility with oil well additives. These are presented in Table 12.

5.4. Hydrogen Sulphide

This gas (H$_2$S) in solution with oil, hydrocarbon gas or condensate produces so-called 'sour' conditions.

There is general acceptance now that nitrile rubber will give problems if used in sour conditions.

The Du Pont test programme, referred to in the previous section (5.3.2)[13] also included exposures to sour conditions which are summarized in Table 13.

At 200°C, as before, only fluoroelastomer was tested, and all results were satisfactory.

Some additional data which include TFE/P have been obtained for sour kerosene containing 12% H$_2$S are given in Table 14. They confirm the poor resistance of nitrile to sour kerosene, but are perhaps unfair because of the high temperature used.

TABLE 13

EXPOSURE OF ELASTOMERS TO SOUR CONDITIONS (THREE DAYS AT 150°C): COMPOUND HARDNESSES 70–80 AND 90–95

Test fluid	Nitrile	FKM A	FKM B	FKM GF
Sour oil[a]	Brittle	Satisfactory	Satisfactory	Satisfactory/good
Sour gas[b]	Brittle	Good/satisfactory	Satisfactory/good	Satisfactory

[a] 90% kerosene, 10% H_2S, 2% H_2O.
[b] $H_2S/CO_2/CH_4$ at 20/5/75 vol % + 2% H_2O.

The superior performance of peroxide-cured Viton® GF is apparent over both standard Viton® AHV and Aflas.

A further study in wet sour oil at high pressure compared TFE/P and FKM GF. The data shown in Fig. 6 confirm that the peroxy-cured Viton® will maintain its original properties better; in particular, less softening or modulus reduction occurs.

Data quoted by Hull[5] for Aflas 150P in 35% wet sour gas at 200°C (Table 15) indicate a somewhat better performance profile for this polymer type. It may be that hot sour oil (Table 14 and Fig. 6) has a more aggressive effect on TFE/propylene polymer than sour gas. Suffice it to say that both fluoroelastomers and TFE polypropylene rubbers are likely to give much better service in sour conditions than nitrile rubber, regardless of temperature. As is the case with corrosion inhibitors, perfluoroelastomer will give maximum resistance to sour conditions.

The data shown in Table 16 were obtained from an actual seal after service at 140°C for 34 months in sour gas. The gas composition was CO_2 15–19%, H_2S 9%.

TABLE 14

COMPARISON OF ELASTOMERS AFTER ONE WEEK AT 200°C IN KEROSENE CONTAINING 12% H_2S

Property retention	FKM GF	FKM AHV	TFE/P	Medium high acrylonitrile
Tensile strength change (%)	−34	−50	−43	−80
Elongation at break change (%)	0	+100	−5	Brittle
Hardness points change (IRHD)	−11	−13	−22	Brittle
Volume swell (%)	2	2	21	1

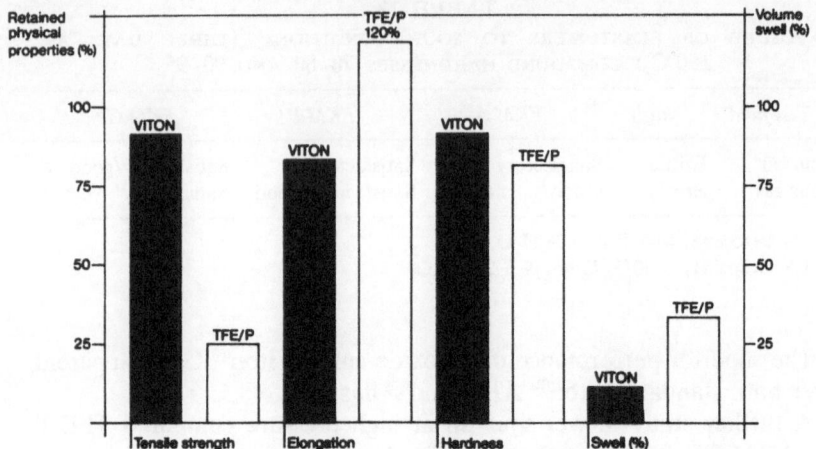

FIG. 6. Resistance of Viton® GF and TFE/P to wet sour soil. The elastomers were exposed to H_2S (11·5%), CO_2 (11·5%), H_2 (38·5%) and SAE 20 oil (38·5%) for six days at 204°C/14 000 psi (97 mPa).

TABLE 15

AFLAS 150P AGED 100 HOURS AT 200°C IN WET SOUR GAS[a]

Property	Original	After ageing
Hardness (Shore A)	83	80
Elongation at break (%)	110	95
Tensile strength (MPa)	18	12·5
Volume swell (%)	—	3

[a] 35% H_2S, 50% CH_4, 15% CO_2, 10% H_2O

TABLE 16

PERFLUOROELASTOMER V-RING PERFORMANCE (140°C/34 MONTHS/SOUR GAS)

Property	Before	After
Tensile strength (MPa)	21	15
Elongation at break (%)	30	50
Hardness (IRHD)	95	95

There have been to date approximately 1600 completions in which Kalrez® has been used in sour conditions without failure.

5.5. Steam and Aqueous Acid Conditions

Resistance to steam and aqueous acidic media in addition to the other previously mentioned oil well environments is necessary. Steam is mandatory for well enhancement and in geothermal conditions. These conditions prove too severe for most medium-rated elastomers.

Fluorinated elastomers are usually looked upon as being best equipped to withstand these conditions. The level of resistance can depend on the degree of fluorination of the backbone of the elastomer and the details of the compounding used. Advances in steam resistance have recently been made with the introduction of high-fluorine peroxide-curable Viton® GF and also peroxide-curable TFE/P Aflas 150P.

Some data extracted from the literature over the range of fluorocarbon elastomers are shown in Table 17.[5,9,14] The clear advantage is the property retention in the compounds of peroxide-cured Viton® GF and Aflas 150P compared with the more conventionally compounded Viton® B candidate. The perfluoroelastomer predictably has little change in its properties.

TABLE 17
STEAM RESISTANCE OF ELASTOMERS

Property	FKM GF	FKM B	TFE/P 100 h/260°C	PFE 14 days/230°C
	6 days/200°C			
Tensile strength change (%)	−52	−91	−35	−10
Elongation at break change (%)	+128	+300	+45	+200
Hardness change (IRHD)	−8	−22	−1	−2
Volume swell (%)	12	8	2	10

5.6. Methanol

The elastomeric seal is required to have the best resistance to the range of organic chemicals which are used in oilfield exploration and production. In addition to the obvious necessity for resistance to crude oil, the elastomer must not be affected by aromatic and alcohol fluids.

Specifically, methanol is an area of possible concern: it is often injected into gas wells and pipelines to prevent hydrate formation.

Experience gained from the observation of oil seals in the automotive industry indicated that blends of methanol and gasoline can have a deleterious effect on many elastomers currently used in automotive oil seals. The effect of methanol as a constituent (15%) in an aromatic fuel is greater than that of other higher alcohols, e.g. ethanol, as is shown in Fig. 7. This should be compared with the swelling behaviour of the same range of elastomers in pure 42% aromatic fuel (Fig. 8).[15]

The strong relationship between swelling in methanol and the bound fluorine level in the fluoroelastomers, seen in these examples, is further accentuated when the equilibrium swell in pure alcohol is examined (Fig. 9).[15] The PFE and TFE/P elastomers both exhibit very good resistance to methanol.

The importance of the correct selection of the elastomer to be used in the seal is clearly indicated. Furthermore the differences that can be obtained between different types of elastomer within a generic family is also evident from these volume swell data (Figs 7, 8, 9).[15] A closer examination of the physical properties on ageing in methanol of two candidate elastomers, FKM B and FKM GF, preferred for their improved methanol resistance, is given in Table 18. This shows significant differences between these two apparently generically similar materials.

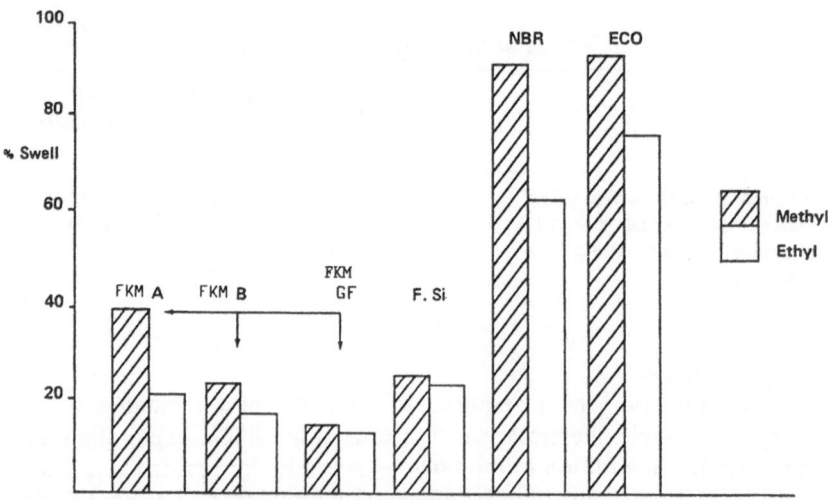

FIG. 7. *Effect of alcohol/gasoline blends on elastomers at 54°C*: the equilibrium swell in 15% alcohol blends (base fluid 42% aromatic) is shown.

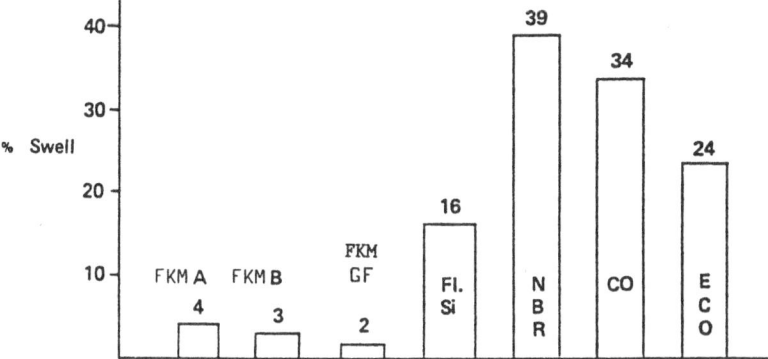

FIG. 8. *Effect of alcohol/gasoline blends on elastomers*: swelling in 42% aromatic fuel at 21°C for 20 days.

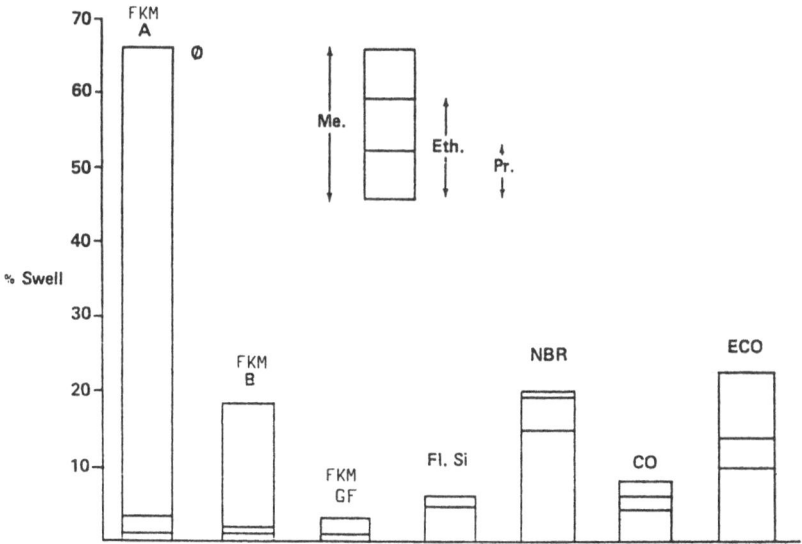

FIG. 9. *Effect of alcohol on elastomer*: swelling in pure alcohols at 21°C for 20 days. Divisions represent swell in ethyl and propyl alcohols. Where only one division is shown, swell in ethyl and propyl alcohols was very similar.

TABLE 18

RESISTANCE TO METHANOL (SEVEN DAYS/METHANOL/24°C)

Property	FKM GF	FKM B
Tensile strength change (%)	−25	−50
Elongation at break change (%)	+4	+165
Hardness change (IRHD)	−2	−27
Volume swell (%)	3	32

5.7. Peroxidized Diesel Oil

Diesel fuel is often used to prepare drilling muds. If stale (peroxid-'ized) diesel is used, there has been concern as to its effect on blow out preventer seals.

Some work carried out by Du Pont has indeed shown that such stale fuel oil has a very deleterious effect on nitrile rubber. Nitrile is now the most common elastomer used to make BOPs.

Table 19 shows this effect, in comparison with the same data obtained for fluoroelastomer, which is obviously superior.

TABLE 19

EFFECT   OF   STALE[a]   DIESEL   OIL   ON   NITRILE   AND FLUOROELASTOMER COMPOUNDS (18 DAYS AT 94°C)

Property	Nitrile	Fluoroelastomer FKM AHV/E45 blend
Tensile strength retained (%)	20	100
Elongation at break retained (%)	20	70
Hardness change (IRHD)	+15	−5
Tear strength retained (%)	40	100

[a] 770 ml Diesel oil + 7 ml t-butyl peroxide to give peroxide number of 80.

6. SPECIFYING THE SEAL ELASTOMER

Any elastomer used in a seal must meet three basic requirements. Firstly, the material must be chemically resistant to the fluids with which it will be in contact. Secondly, it must be able to withstand the temperatures involved both in operation and during storage. Finally, the material must be able to stand up to the mechanical environment, including such factors as pressure, stressing and abrasion.

These requirements are interdependent and a change to improve one property may result in the deterioration of another. Properties of elastomeric parts are strongly influenced by the selection of ingredients, fillers, etc. and other aspects of compounding.

6.1. Specifying for Automotive Applications

In order to have the desired performance from the elastomeric seal it is necessary to specify performance properties which the seal must meet, over and above the basic selection of an elastomer type.

The automotive industry was confronted with this same problem, namely to define the performance requirements of seals, such that the wide range of possible environmental conditions could be reflected in the test regime.

The American Society for Testing Materials (ASTM) has issued a joint report covering a classification system for rubber materials for automotive applications[2] which is based on performance rather than chemical composition. The classification system is based on the premise that the properties of all rubber products can be defined by 'types', based on resistance to heat ageing, and 'classes', based on resistance to swelling in a standard oil (ASTM No. 3). The complete specification is, of course, more complicated than this and results in what is known as a 'line call out' as shown in Fig. 10.

As an example, a line call out M6HK 810 A1-11 B38 EF31 can be interpreted as follows.

M indicates that the International System of Units (SI) is being used. The prefix 6 indicates the grade number and that the material must meet more than just the basic requirements. The grade number tells the rubber chemist which set of additional requirements is needed and these are defined in the specification.

The letter H designates the 'type' which, in this case, must be capable of withstanding 250°C with a maximum change in tensile strength of ±30%, an elongation change of −50% maximum and a hardness change of ±15 points. The letter K indicates the 'class' which, in this case, calls for an elastomer that does not swell more than +10% when aged in ASTM No. 3 oil for 70 h at 150°C.

The first digit 8 of the three-digit number sequence is the hardness requirement and means that the hardness should be 80 ± 5 Shore A

<u>M6 HK 810 A1–11 B38 EF31</u>

(Line call out)

Basic Requirements

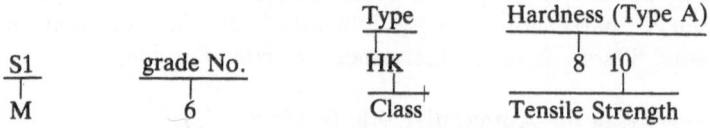

Suffix Requirements

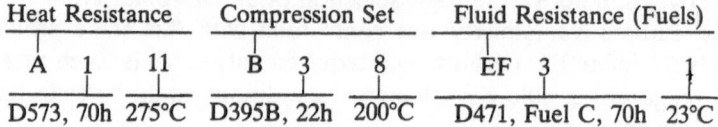

FIG. 10. SAE J200/ASTM D2000 'line call out'.

points. The second and third digits, 1 and 0, represent the minimum tensile strength required, in this case 10 MPa. These complete the basic requirements of the 'line call out' except for the compression set requirement of 35% maximum when tested for 22 h at 175°C. Compression set is not included as part of the 'line call out', but is one of the basic requirements for all cases and is specified in J200/D2000 for each combination of 'type' and 'class'.

The remaining sets of letters and numbers are called suffixes and further refine and define the properties required. A1-11 narrows the amount of change in the tensile properties after heat ageing for 70 h at 275°C. B38 limits the amount of compression set which can be tolerated after 22 h at 200°C to a maximum of 15% and EF31 limits the amount of swell and change in properties that can be tolerated after ageing 70 h in ASTM Reference Fuel C at 23°C.

Hence it is possible first to identify the class of polymer and then to establish its basic mechanical properties. Additional requirements can be specified by referring to tables in the specification which concern test methods, fluid types, temperatures, etc.

6.2. Specifying for Oil Industry Use

The needs of the oil industry are different from those in the automotive industry but it is most important that the oilfield engineer obtains the correct sealing material for the job in question. In order to ensure that the correct material be selected and relevant seal properties be identified and met, a selection or call out system specific to oilfield conditions was proposed.[16]

Such a system could be developed initially along the general lines of the ASTM D2000 system described above, with initial screening and positioning of the elastomeric material being made on the grounds of heat resistance and oil resistance.

A further refinement of such a classification system has been put forward based on the compatibility between elastomeric materials and fluids as set out in ISO/DIS.6072.[19] This proposal by Nagdi is based on the swelling behaviour of an elastomer in specified fluids, relating the swell results to those of a known standard compound. The refinement in this method means that more precise definition of the elastomers being used in a seal is made, the polymer base being identified.

The more specific property requirements relevant to oilfield usage would then be defined by separate suffixes. The tests envisaged would address the problem areas for seals discussed already in this chapter.

Typical specific tests which could be envisaged are:

Explosive decompression: repeated cycling of set conditions on a test rig of the type described by Cox.[8] Temperature, gas compositions and pressure cycles would be further defined (Section 5.1 this chapter).

Extrusion resistance: a test similar to the API test described earlier in this chapter (Section 5.2).

Corrosion inhibitors: industry-wide agreed testing in NACE A and NACE B containing system as described in Section 5.3.1.

Hydrogen sulphide: Ageing evaluation in synthetic 'sour' media as described in Section 5.4.

Steam resistance and methanol resistance: could be defined by simple screening tests (Sections 5.5 and 5.6).

When this idea was proposed at the conference *Offshore Engineering with Elastomers* in Aberdeen (1985)[16] it raised considerable interest from all quarters of industry concerned with the elastomeric seals: the oil companies, oil tool companies, rubber part/seal manufacturers and the raw material suppliers. Such a system, where the specification could be based on relevant tests, was welcomed, although the setting up of such a test regime with industry-wide agreement would require some considerable effort.

The individual competitiveness of one manufacturer over another is not lost, since special expertise can be offered over and above the general guidelines.

The advantage of such a system is that a user can have confidence that he will receive seals conforming to an agreed minimum standard.

REFERENCES

1. Bowyer, M. L., 'Application and selection of elastomers for downhole equipment in oil and gas wells', *Rubber in Offshore Engineering Conf. Proc.*, April 1983.
2. ASTM D2000 Classification System, *Annual Book of ASTM Standards*, Part 37, 483.
3. *Vanderbilt Handbook*, R. T. Vanderbilt Inc., Norwalk, USA, 1978, p. 171.
4. Sweet, G. C., 'Special purpose elastomers', in *Developments in Rubber Technology—1*, ed. A. Whelan and K. S. Lee, Applied Science Publishers, 1979, Chapter 2.
5. Hull, D. E., 'Oilfield media profile of TFE/propylene co-polymer', *124th ACS Rubber Div. Meeting*, Xenox Inc., Houston, USA, Oct. 1983.
6. Ender, D. H., *Swelling of some Oilfield Elastomers in CO_2, H_2S, and Methane at Pressures of 28 mPa*, Shell Development, Westhollow Research Center, Houston, USA.
7. Potts, D. L., 'Explosive decompression of elastomers in gas duties', PRI Conf. *Offshore Engineering with Elastomers*, Aberdeen, June 1985, Paper 21.
8. Cox, V. A., 'Service failures—a user view', PRI Conf. *Offshore Engineering with Elastomers*, Aberdeen, June 1985, Paper No. 19.
9. Weston, R. J., 'High performance elastomers for seal materials', *Rubber in Offshore Engineering Conf. Proc.*, April 1983.
10. *Valve Report*, Circular PS 1134-46-72, American Petroleum Institute, 1958.
11. Stevens, R. D., 'Compounding fluoroelastomers for resistance to extrusion at high temperatures and pressures', *Kautschuk u. Gummi Kunstoffe*, **37**(9), 1984.
12. Watkins, M. J., 'Watch out for elastomer inhibitor incompatability', *Petroleum Engineer*, April 1984, 28.
13. Pugh, T. L., 'Evaluation of fluoroelastomers for oilfield service', *ACS Rubber Div. Meeting*, Oct. 1983.
14. Du Pont Bulletin, *Viton® GF (VT.250.GF)*, E. I. Du Pont de Nemours, 1981.
15. Nersasian, A., MacLachlan, J. D. and Brown, J. H., *Resistance of Elastomers to Fuels Containing Alcohols*, SITEV, Geneva, May 1981.

16. REVOLTA, W. N. K. and SWEET, G. C., 'A review of elastomers, used in oil field environments', PRI Conf. *Offshore Engineering with Elastomers,* Aberdeen, June 1985, Paper No. 16.
17. BAKER, G. R., *A Primer of Oil Well Drilling,* Petroleum Extension Service, University of Texas, Austin, Texas, USA, 1979.
18. BAKER, G. R., *Oilwell Service and Workover,* Petroleum Extension Service, University of Texas, Austin, Texas, USA, 1979.
19. NAGDI, K., PRI Conf. *Polymers in Oil Exploration Conference,* London, 12 June 1986, Paper No. 10.

Chapter 6

USING MODERN MILL ROOM EQUIPMENT

H. ELLWOOD

Farrel Bridge Ltd, Rochdale, Lancashire, UK

1. INTRODUCTION

The following chapter has been written to outline some of the general principles involved in the specification, planning and operation of modern mill room equipment.

As soon as one problem is solved then another arises; each new step forward leaves a footprint but also raises a cloud of dust.

The apparent changes are small and tyres are still round and black with a hole in the middle—but a whole new technology has gone into the design, formulation and performance of the radial tyre as compared with the cross-ply. Machinery which was perfected for the high-volume cross-ply tyre production is not necessarily best for radial tyre making.

Hose pipes are still long and flexible but the service requirements (pressure, life, temperature and contact liquids) are constantly changing.

The challenge has had to be faced of making large flexible electrical cables, for modern dockside and mining equipment, with very arduous duties and high current capacities.

Oil seals and automotive parts now have to meet much higher requirements and the 200 000-mile car engine is just around the corner.

This chapter was written in Singapore when the author was on his way home from working on a very new continuous rubber mixing plant. Most of the problems encountered were associated with the climate and environment.

Electrical, electronic and computer equipment gave problems due to severe electrical storms. Extra cooling water was required to meet the high ambient temperatures. Dust emission from the rubber granulator and transfer bins had to be stopped completely as the pre-blend plant for the continuous mixer was located in the middle of the extrusion shop. When a similar new plant is envisaged these problems will be addressed at the planning stage.

The training of labour to use new equipment is very important and the costs of training must form part of the project. Explaining new methods of operation to rubber workers with years of experience using existing equipment is of vital importance, as is listening to their views on the best way to make a product.

The author has been to plants where the formulas have been adjusted by the operatives to make the product workable on the old equipment, so when new equipment is used and the 'official formula' put in, the results differ widely from the product normally produced. Training and communication are therefore extremely important.

2. STOCK PREPARATION

2.1. Uniformity

Uniformity of production is important to achieve the design properties of the product and to ensure the processing of materials successfully with minimum variation and scrap.

Unfortunately the raw materials used in compounding are, as in any other production, not uniform in all respects. Materials from different sources can have slightly changed properties, rubbers, carbon blacks and fillers being particularly subject to variation.

The best way to achieve better uniformity is to blend the materials as thoroughly as possible. For example, in large-scale production it is best to store carbon blacks in at least two storage silos for each grade and to make weighings by using some from each of two shipments. In the case of rubber this would mean one bale from each of two pallets.

The 'small chemicals' have the biggest effect on the formula's cured properties. Uniformity of the supplies of sulphur and zinc oxide should be carefully monitored. Mineral fillers are very much source-dependent as to their properties: a material from one source can easily be processed whilst the apparently identical material from a different source can be highly abrasive and hard to process.

Powder which will be effectively conveyed and handled from one source can be completely different to deal with from another source. Care must be taken in selecting the equipment to be used if materials are to be obtained from different sources.

In monitoring uniformity is is important to consider the method of shipment and packing. Rubbers are particularly difficult to handle due to a wide variety of packing methods. Some are wrapped in polyethylene sheet which has a low melting point and which can in most cases be mixed into the compound; others are packed in paper sacks or tough plastics which must be completely removed.

2.2. Preheating and Preconditioning Raw Materials

In the early days of rubber compounding it was general practice to preheat natural rubber bales before processing. This practice has been largely discontinued in recent times. This is due to the increased use of synthetic rubbers, and the higher installed power of modern mill room equipment which can now process natural rubbers that have become crystalline due to low temperatures (frozen rubber).

Some synthetic rubbers are also subject to crystallization due to low temperatures in storage and shipment. Induction heaters have been used with success in preheating bales to unfreeze the rubber.

The effects of mixing soft compounds using rubbers which are frozen is that small nibs of unmixed polymer which can spoil an extruded section are left in the compound. In harder compounds (*ca* 70 ML1 + 4 Mooney viscosity at 100°C) this is less of a problem, but variations in the temperature and properties of the polymer can affect the mixing cycle and lead to variability of the mixed compounds.

In a similar manner the moisture content of the raw materials should be carefully monitored. Some materials are hydroscopic and can take up water very easily. Fibrous fillers are particularly prone to water take-up.

The use of air as a conveying medium can, in certain climatic conditions, cause problems as the moisture content leads to difficulties in handling and changes in the mixing time, and it can also make the mixed compound porous.

Storage conditions for rubber chemicals should be dry and warm as such materials can be damaged, with great financial loss, by unsuitable warehousing.

2.3. Weighing Accuracy

Inaccurate weighing is the biggest single contributor to variations in mixed compound properties.

Rubber chemists often note that the compound made under laboratory conditions differs from a batch made as normal factory production. Whilst laboratory mixers do, as a general rule, impart more mixing per unit volume than a production-size mixer, the laboratory mixer giving more intensive mixing for the same temperature rise, it is also very likely that the chemicals are far more accurately weighed under laboratory conditions.

Of course weighing is only part of the process in getting the correct proportions of chemicals into a mix. In automatic weighing, chemicals can lodge and later dislodge in the chutes and conveyors to give variability of the chemicals in the mix. Chemicals can be spilt or lost into a dust collection system.

A typical effect of variation in cured properties is shown in Fig. 1.

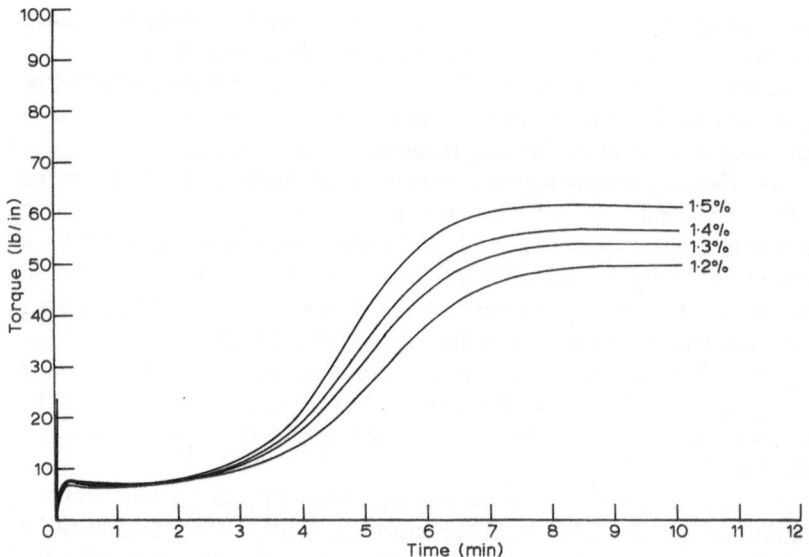

FIG. 1. Standard Monsanto rheometer chart for stock mixed on mill with various proportions (1·2–1·5 wt%) of the sulphur/accelerator curing system. Conditions: chart motor, 6 min; range sel., 50; arc ±1; temperature, 180°C.
© Farrel Bridge Ltd, 1987.

Here the amount of sulphur/accelerator was changed slightly from 1·2 parts to 1·5 pph. In a batch size of 50 kg this means a variation of sulphur/accelerator of 150 g will produce this range of variability. So the odd little spillage of sulphur on the floor or sucked up into the dust hood can greatly affect the product quality.

With weigh scales it is important to consider very carefully the accuracy of the scale. If the scales are linked to a computer then the resolution of the analogue/digital converter, which controls the signal into the program, is important. Some A-to-D converters have only an eight-byte capacity and divide the full scale deflection of the weigh scale into only 256 steps, i.e. in a 0–50 kg scale into 180-g steps. At least 12 bytes should be used to get 2048 steps (24-g).

The construction of chutes, valves and conveyors should be such that they are self-cleaning, if possible arranged so that one of the other components in the mix cleans out any residue; for example in an operation where pellet rubber is used, this is ideal to follow a sulphur/accelerator feed down a chute to sweep it clean. Vulcanized rubber sheet makes an excellent material for the construction of chutes.

A mixing plant with a large amount of powder/oils on the floor and covering the machinery is a good indication that not all the chemicals intended to reach the mixer have done so and, as a result, the mixed compound will not be very consistent.

2.4. Polymer Blends

If polymers, for example SBR and natural rubbers, are to be blended, the initial 'elastic properties' should be considered.

Unmasticated natural rubber is much 'stronger' and can be mechanically deformed in the mixer without doing any permanent change to the shape or properties, if the rubber is in pieces small enough to pass through the clearances in the mixing chamber. This means that the size of the rubber to be fed into the mixer is important when blending polymers of varying hardnesses and elastic properties.

The best demonstration of this is that if vulcanized hard rubber balls of diameter about 1 in (25 mm) are put into a rubber mixer along with the rest of the mix to check, for example, 'dispersive mixing', the majority of the balls will be undamaged as the mixer is discharged.

It is best to feed bales of unmasticated natural rubber with synthetic rubbers into the mixer so that the rubbers are deformed and heated to soften them, then they will blend together.

If feeding granulated natural rubber, then the pieces should be at least 40–60 mm across, so the mixing rotors can 'bite' on the rubber as it is fed into the mixer. It is, however, possible to feed small pieces of unmasticated rubber into a continuous mixer and get a good dispersion (see continuous mixing—Section 7).

2.5. Thermoplastic Rubbers
The Banbury mixer is ideally suited to the making of thermoplastic rubbers. Polystyrene/SBR blends can be mixed in 3–4 min and TPEs requiring higher temperatures in 5–6 min. The requirement is to have a mixer with high rotor speeds, e.g. 120 rpm for an 80-litre mixer, discharging into an extruder with an underwater pelletizing head.

The Banbury must be heated (a temperature range of 40–150°C has been found to be suitable) and the extruder must also be heated to 150–180°C.

Process oils have to be added to the mix at about 150°C, which can be dangerous with the risk of an explosion. It is best, therefore, to use an oil injector which can pump the oil directly into the mixing chamber with the ram of the mixer down to avoid contact with air.

It is important to mix at a sufficiently high temperature to melt and mix the thermoplastic completely—several lifts of the ram should be made to ensure all the thermoplastic is within the mass of the mix to enable it to be melted.

A mill can be used to accept the batch from the mixer and then the thermoplastic rubber can be cut and rolled on the mill in the normal manner.

3. BATCH MIXING

3.1. Machine Preheating
The mixing of rubber uses electrical energy which is turned into heat. The mixing/milling/extruding machinery is heated up by the work done by the sliding of the rubber on the metal surfaces, and also by thermal conduction from the rubber to the metal. The metal has to reach a surface temperature to give a sufficient temperature gradient so that the heat energy of the rubber transmitted to the surface of the metal can flow by conduction through the metal and be transferred to the cooling water.

Typical temperature differences from a two-roll mill are:

Rubber 90°C
Roll outer surface 30°C
Roll internal surface 20°C
Water 15°C

However, at the start of a production run the machinery is cold so it is necessary to heat the mass of metal of the machinery. With a two-roll mill this is normally done by starting to mix with the cooling water supply turned off, thus allowing the metal of the rolls to be heated. This mix will, of course, be different from the next batches as the temperature at which the rubber is processed is different. The amount of energy put into the mix will vary because of this and it can show up in final uniformity variations.   ·

Rubber is a poor transmitter of heat. The most dramatic example of this is the way in which it is possible to leave a melted streak of rubber from a car tyre on a very cold road surface if the brakes are applied hard, changing the kinetic energy stored in the mass of the car into heat energy at the point of contact between the tyre and the road. The speed of heat build-up in the surface layer of the tyre is so fast that the rubber cannot take away the heat and so the friction heats up the rubber to approximately 200–400°C to melt the rubber.

Another example which has caused some problems in the past is when a mixer discharges its batch during the initial feeding stages with the ram still up. This is caused by the door top thermocouple tip being locally heated to a very high temperature as the cold rubber being fed into the chamber skids over it—sending a signal to the temperature-actuated automatic discharge instruments. The solution to this problem is to fit a timer to inhibit the automatic discharge instrument until the ram has been lowered.

In rubber processing local high temperatures which can be generated by rubber skidding across cold metal must be avoided. In order to start a mixing production run it is an advantage to preheat the mass of the machine metal to the temperature it will achieve during a production run. The Banbury mixer with its big batch size, due to the large fill factor, needs three or four batches to achieve temperature equilibrium if no preheat is used; the intermeshing rotor machines, such as the Shaw Intermix, have a greater metal mass compared with the batch weight and require five to seven batches before heat stability is achieved.

TABLE 1

TYPICAL MACHINE TEMPERATURES (°C)

Stock	Sides	Rotors	Top
Natural rubber mastication	30	30	40
Natural and synthetic rubber and blacks	30–50	30–50	40–80
EPDM	50–60	50–60	50–60
Thermoplastic rubbers	40–150	40–80	40–80

For 'normal mixing' preheat temperatures of 30–40°C are sufficient but for some stocks which stick to cold metals (EPDM, etc.) 50–60°C are good starting temperatures. The transfer of heat is time-related, i.e. the shorter the mixing cycle the less the heat transfer from rubber to metal and from metal to water.

Table 1 lists typical operating conditions giving suggested circulating water temperatures for the mixing chamber sides, mixing rotors and discharge door top.

3.2. Ram Pressures

Banbury and other internal mixers are equipped with a ram, actuated by an air cylinder, to feed the material into the mixing chamber during the mixing cycle.

Visco-elastic materials, unlike liquid or paste-like materials, need to be restrained and the ram is used to prevent some, or indeed all, of the mix working its way out of the chamber up the feed throat.

The question is, how hard to push the material into the chamber?

Movement of the ram is necessary to prevent overloading of the drive motor. Many years ago the author went to see a 3A Banbury, used to mix hard ebonite stocks, because the motor driving the machine kept going into overload and stalling on every batch, although the machine had been used to mix the same compound easily for many years without overload. The fault was traced to a crack in the hopper cover which trapped the piston rod as the force from the mix thrust upwards, preventing movement of the ram. This caused the overload condition.

This example is cited to show how it is possible to control the load on the mixer motor by regulation of the air pressure. In fact it is a good thing to arrange a system to relieve air pressure as the load on the motor reaches an overload condition.

Let us return to the question of how much pressure should be used.

(i) Sufficient pressure is needed to push the batch into the mixer at the start of the cycle. Lumpy rubber can get trapped between the ram and the feeding throat, so to push down at full pressure can mean that the ram is stuck and cannot be pulled out of the mix. Therefore, the use of just below full air pressure is best.

(ii) The pressure used should enable the practical maximum batch size to be obtained. As the mixed materials metamorphose from elastic lumps, powders and oils into a homogeneous plastic mass the amount of material which can be contained in the mixing chamber changes. It is necessary for the ram to reach a low position early in the cycle as material under the ram in the feed throat is not being, and will not be, mixed until it gets down into the chamber.

The position of the ram relative to the mixing rotors and chamber in the down, or near-down, position affects the 'between rotor' mixing action. A large mass of stationary rubber under the ram prevents the proper cross-blending of the rubbers.

In the initial use of 'low-pressure ram' low-power Banbury mixers it was necessary to add the rubber and chemicals slowly. The ram was lifted many times and this allowed the mix to turn over as the ram was lifted to allow good cross-blending to take place. In modern high-pressure/high-power mixers with a 'one-shot feed' of rubber and chemicals a reduction or air pressure during the cycle, to allow the ram to move up the feed throat 200–500 mm for 5–10 s during the mixing cycle, can greatly improve the dispersive distribution of the mix.

Ram shape and clearance modification can help to enhance the performance of a mixer, particularly on machines dedicated to a restricted mixing duty. This should be discussed with the mixer manufacturer to get the best from the equipment.

The speed of ram movement should be as fast as possible, but very fast movement can result in excessive dust being raised as the ram moves, thus displacing the air in the chamber. Some powder will flow very easily when aerated. This aeration is caused when the powder falls into the mixing chamber. If the ram is moved very quickly the powder flows up over the ram and further ram lifts are required to get the material down into the chamber. Reduced ram pressure can help with such materials as the ram will move up and down more at the low pressure by the interaction between the rotors and the stock, shaking the material from above the ram down into the mixing chamber.

3.3. Rotor Speeds

To achieve the required degree of mixing a certain number of rotor revolutions are required; i.e. it does not depend on mixing time or energy input, even if these two convenient methods of counting rotor revolutions (time) or measuring how much energy has been used in the machine turning the rotors a certain number of revolutions.

While the rotors are turning the temperature of the mixed material usually increases. The rate of temperature increase is determined by the rate of rotor rotation (rpm). The faster the rotor moves the faster is the rate of energy input and the quicker the temperature rise.

The cooling of the rubber from the mixer metal (rate of heat transferred) depends on the thermal conductivity of the material of construction (K), the surface area (A), the thickness of the metal (h), the temperature difference (ΔT) and the time (t):

$$\text{Amount of heat transfer} = \frac{K \times A \times \Delta T \times t}{h}$$

As K, A and h are fixed within one machine design the only variables available to the machine user are ΔT and time. However, ΔT (the difference between the rubber/metal/water temperature) is variable, i.e. at the start of the cycle the metal of the mixer will be hotter than the rubber and only towards the end will the rubber be higher in temperature than the metal, so the use of ΔT is not of very much help. The time is therefore the biggest factor in the exchange of heat energy from the mix. Therefore:

To reduce both time and total energy input, increase rotor RPM;
To increase both time and total energy input, reduce rotor RPM.

To achieve the necessary rotor revolutions to get the required dispersion/viscosity reduction, it may be necessary to do the mix in several stages to control the temperature rise.

3.3.1. Energy Inputs

Typical energy inputs for a two-stage natural rubber mix in a 40-litre Banbury mixer are as follows.

Masterbatch mix time 2·2 min total (for feed and discharge 0·5 min and 1·7 min for mixing). At 70 rpm the rotors turned 119 times in the 1·7 min.

Final mix time 2·25 min total. Again 0·5 min was used for feed and discharge, and 1·75 min for mixing. At 30 rpm the rotors turned 53 times in the 1·75 min.

Therefore a total of 119 + 53 = 172 revolutions were used for both stages of the mix.

The energy was used as shown in Table 2. The total energy used for both stages was 0·2255 kWh/kg.

With the introduction of variable-speed DC drives to batch mixers it is now possible to change the rotational speed (rpm) of the rotors during the mixing cycle.

3.3.2. Variable-Speed Mixing

It could be an advantage to start a mix at a fast rotor speed until the batch reaches a temperature high enough to allow energy to be transmitted through the metal to the cooling water, and then to slow down the rotational speed to slow down the rate of temperature rise to allow more energy to be put into the batch.

Slowing down the rotors can change the indicated temperature of the thermocouple. However, unless the rotors are run very slowly it would be unusual for the actual rubber mix to reduce in temperature. For example, if a mix has to be taken up to 160°C in the first stage, reducing the rotor speed to allow the batch to cool before adding the sulphur is in general not a practical proposition.

When feeding hard rubbers into the mixer a slower rotor speed can

TABLE 2

ENERGY USED (kWh/kg) IN TWO-STAGE PROCESS IN A 40-litre
BANBURY MIXER

	Masterbatch	Final mix
Energy to heat rubber	0·0814 (56%)	0·0508 (62%)
Energy used electrical and mechanical	0·0129 (9%)	0·00737 (9%)
Energy to cooling water	0·0493 (35%)	0·02372 (29%)
Total energy used	0·1436 (100%)	0·0819 (100%)

reduce the electrical loads at the start of the cycle. The rotors can be speeded up as the mixed material is softened.

As the harder stocks are generally more 'scorchy', because they increase in temperature more rapidly when being mixed, lower speeds are used. To get increased torque at lower speeds very often a DC motor which is 'slant rated' using 'field weakening' is specified. What this means is that it is possible to specify a DC electric motor so that it develops its full horsepower at 80% full speed as well as at full speed, e.g. 100 hp at 1200 rpm and 100 hp at 1500 rpm. This means that the torque up to 1200 rpm is constant, and it is reduced between 1200 and 1500 rpm. The 'slant rated' or 'field weakened' specification is ideal for a mixer drive, so as to have high torque at low speeds and reduced torque at high speeds.

Two-speed gearboxes are another way of increasing the torque at lower speeds.

3.4. Feeding Techniques

(i)   *One-shot feed*: polymer immediately followed by powders and the ram down—oil injected through mixer end frame or through the mixing chamber side.

(ii) *Upside-down feed*: powders first and then polymer—then ram down—oil injected with ram down.

(iii) *Sequential feed*: first rubbers then ram down—ram up followed by some or all powders and ram down. Oils can be poured down the throat with ram lifted.

(iv) *Over-the-ram feed*: first rubbers then ram down—powder fed in with ram down. The bumping of the ram shakes the powders down into the rubber. (This is no longer used except for very old low-power mixers.)

The 'one-shot feed' (i) is the normal masterbatch method with high-power/high-pressure Banbury mixers. The full chamber uses the high power available; very high mixer loads are used which result in good dispersion and machine utilization. Cycle times of $1\frac{1}{2}$–3 min are common with this method.

'Upside-down feed' (ii) is also used for high-power machines; the carbon black can go down into the chamber and be held there by the rubber, as the rubber is pushed down the mixer throat. The carbon black can be compressed by the action of the rotors, in the absence of the rubber, and can be agglomerated. An engineering disadvantage of this technique is that the very high loads caused by compressing the carbon black and having to seal the dry powder in the chamber put a big strain on the dust seals. However, if a large amount of oil has to be put in the mix (typically with EPDM), then putting the powders and oils in first, so that the oil is absorbed into the powders, followed by the polymer, can give good results and keeps the feed hopper and ram clean.

'Sequential feed' (iii) is normally to charge the polymers first to allow them to blend, natural and SBR being typical, ram down then ram lift to feed the powders. This method reduces the load in the mixer and allows lower-powered mixers to mix tough materials. Typical time cycles by this method are in the order of 4–8 min.

The feeding of the curing system (sulphur and accelerators, etc.) is best done early in the cycle. Adding the curing system at the 'last gasp' in the cycle and relying on the mill to disperse what appears to be a 'long yellow streak' can give quality and consistency variations. Some of the rubber is exposed to very high concentrations of the cure system by this method.

When final mixing masterbatches, the addition of the cure system is best if it can be sandwiched in between the masterbatch: i.e. feed half the masterbatch then the cure system, followed by the rest of the masterbatch.

When final mixing masterbatch pellets, a slight delay in sending down the ram can enable the mixer to distribute the cure system before massing together the pellets.

3.5. Temperature Measurement

Now that digital temperature measuring instruments are available it is possible to get accurate readings of the temperature of the thermocouple probe. However, the temperature of the probe is not necessarily the temperature of the rubber.

The temperature the probe measures is some function of

(i)   the temperature of the stock,
(ii)   the thermal mass of the probe,
(iii)   the thermal conduction along the probe,
(iv)   the frictional heat generated by pushing the probe into the rubber, and
(v)   the time taken to do the measurement (the rubber is cooling or heating as the reading is taken).

The thermocouple probes used in internal mixers must be strong and therefore the effects of the thermal mass (ii) and the conduction (iii) can be very high; also the frictional heat will depend on the speed of the mixer. This is often the reason why the apparent temperature of the stock rises quickly. At fast rotor speeds it is just the thermocouple tips being subjected to high frictional forces which causes the rise.

Therefore, the discharge temperature should always be checked against the actual temperature for a specific rotor speed/compound formulation/machine temperature and then the temperature indicator on the mixer should be used as an indication of temperature change and not of the actual temperature of the stock. The same comment applies to infra-red temperature measurement as the infra-red 'emissivity' of the stock depends on the colour, texture, etc., of the stock.

The stock temperature is usually within 0–12°C of the actual dump temperature. If the difference is greater then there is likely to be some fault in the system, for example

(i) wrong type of connection wires,
(ii) crossed leads, or
(iii) wrong thermocouple projection into the mixing chamber.

4. BATCH MIXING PROBLEMS

4.1. Lumps

Lumps of undispersed polymer which can be seen as the batch is discharged onto the mill can be a problem. Lumps of undispersed EPDM, for example, can be disintegrated into 'crumbs' on the mill and will spoil the surface finish of an extrudate. Lumps of SBR and natural rubber can often be mixed out on the mill but can still present a problem.

Lumps can be formed by

(i) using frozen rubbers,
(ii) wrong batch size,
(iii) addition of oil at the wrong time, i.e. too early, allowing the rubber to slip and not be mixed,
(iv) damaged or worn mixer sides, door top, rotor, feed mechanism or ram; broken edges causing cavities; worn doors with too much clearance,

(v)   wrong mixer temperatures—usually the mixer is too cold (see Section 3.1),

(vi)   insufficient or variable air pressure—sometimes a demand for air by another part of the factory can result in 'low air pressure' at the mixer which can result in the ram not reaching the bottom of its stroke on occasional batches, giving a random occurrence of 'lumps', and

(vii)   incorrect operation of the discharge door—mixing with the door partly open can create cavities at the edge of the door (the operation of the latch and the setting of the limit switches should be checked).

4.2. Dispersion of Carbon Blacks and Powders

Undispersed agglomerates of powders can be caused by

(i)   problems in the powders as supplied,

(ii)   moisture, either in the material as supplied or due to storage or handling problems,

(iii)   the mixed batch being too large, and

(iv)   the mixer being worn and needing a change of rotors (NB. Rotor shape is just as important as the machine clearances).

The ability of the mixer to break up agglomerates of powders depends on the viscosity of the mix. It is the mix itself which must break up any unmixed powders, so it is important to keep the mix as stiff as possible for as long as possible. The time when the softeners are added should be carefully considered and be as late as possible in the cycle. An oil injection system which can add oil to the mix with the ram down is of great advantage in getting good dispersion.

4.3. Feeding

Only material which is in the mixing chamber will be mixed! Materials 'over the ram' or 'under the ram' but not in the active part of the chamber will not be mixed, neither will materials stuck in feed chutes, etc.

Materials may finish up 'over the ram' due to

(a)   excessive aeration of the powder in chutes and conveyor systems, and

(b)   incorrect handling of carbon black, breaking up the 'black pelletization' and changing the bulk density of the carbon black which, if badly handled, can be as low as $21 \, \text{kg/m}^3$. This makes it difficult to get the correct weight into the mixing chamber.

Materials may finish up unmixed 'under the ram' due to

(a) incorrect batch size,
(b) too low an air pressure,
(c) feeding slab stock too slowly; the last piece folds under the ram and sits on top of the material in the chamber,
(d) the ram sticking in the feed throat before the bottom of the stroke because of material wedging between the ram and the feed throat (a check should be made for wear on the ram and throat misalignment of the hopper/neck extension to the mixing chamber), and
(e) wear or leakage in the ram air cylinder. Most of the wear in the bore of the cylinder takes place in the bottom part of the cylinder. Excessive wear can cause air leakage at the bottom of the stroke.

Materials may 'stick' in the feeding systems between weigh scales and the mixing chamber. The effect of this can be wild variations in material properties (see Section 2.3).

4.4. Mixer Discharging

It is important that the whole of the mix is discharged quickly and cleanly.

(1) The time for a mixer to discharge depends on the rotational speed of the rotors, i.e. a Banbury mixer at 30 rpm takes twice as long as one working at 60 rpm; normally four to six rotor revolutions are required, i.e. 8–12 s at 30 rpm and 4–6 s at 60 rpm after the door has opened. With a two-speed mixer the discharge time should be changed to suit each speed or set to suit the lowest speed.

(2) Sticking of material to the rotors and or discharge door can be overcome by controlling the temperatures of the metal surfaces. The metal temperature of the rotors in high-speed mixers (60–100 rpm) can be much higher than one would expect from the water temperatures. At high speeds the centrifugal force in the water can hold it in the outer parts of the cooling cavities, behind the rotor tips, so it is not circulating properly. Modification to the method of cooling the rotors has on several occasions solved a 'rotor sticking problem.'

(3) Slow operation of the door hydraulic system can lead to stock being extruded into the gaps of a slowly opening door. This can give door closing problems.

(4) Generally a hotter discharge door can give a cleaner discharge of sticky stocks.

(5) Incomplete discharge can lead to bad mixing on the next batch: as the batch will be oversize, the ram will not reach the bottom of its stroke properly and it will leave unmixed materials.

(6) The discharge should be clean and not dusty. A 'dirty discharge' from a mixer which has normally been clean is very often caused by the ram going too far down into the mixing chamber. A loose piston or worn sides can give this effect. The down position of the ram should be checked to ensure that it is at least 5–25 mm up from the edge of the bore (x on Fig. 2).

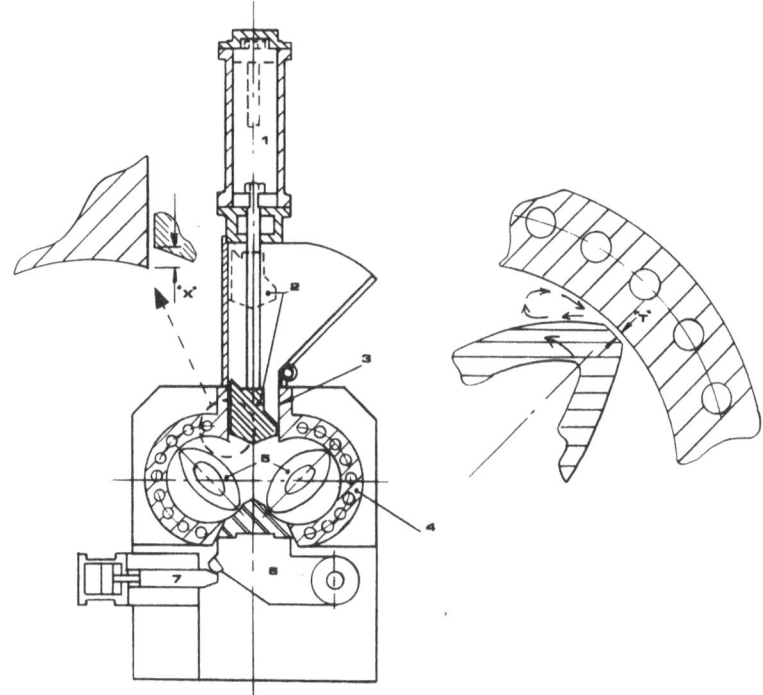

FIG. 2. Section through a typical batch mixer. 1, Ram cylinder; 2, ram; 3, feed throat; 4, mixing chamber side; 5, mixing rotor; 6, discharge door (door top); 7, latch. © Farrel Bridge Ltd, 1987.

(7) To discharge, the ram should be put into 'float position', i.e. downward air pressure exhausted (or balanced by air under piston) before the discharge door is opened. The discharge door should then be closed before lifting the ram.

(8) If stocks stick to the discharge door the formula being mixed should first be examined. Stocks which contain resins and other tackifiers are usually those which can cause sticking. These resins should not be loaded in plastic bags or in one concentrated lot but spread out so they are not fed in such high concentration.

4.5. Uniformity

Whilst the use of mills following the mixer can usually smooth out any uniformity variation, ideally the batch at discharge should be generally uniform.

As stated in Section 2.3, accuracy of weighing is the biggest single contributor to uniformity, but the temperature of the feed stock, consistency of air pressure, machine temperature and setting all can contribute to improve the uniformity.

The most significant step in recent times to uniformity and control of uniformity has been the introduction of computer control and data acquisition in mixing plants.

Whilst a mixer cycle can be controlled by the use of simple hand-set timers, a power integrator and temperature air pressure and speed settings, the ability to select repeatedly the same setting and then report by a computer print-out makes the use of a computer really worthwhile.

5. POST MIXER, SHAPE AND COOL

The batch mixer discharges a large hot lump of mixed material. In this form it has a large volume but low surface area. If left in the 'as discharged' form it would take a very long time to cool and be difficult to process further.

It is therefore necessary to change the shape of the material discharged from the mixer so that it can be cooled quickly and further processed.

Three options are possible:

(a) to shape the material into thin sheet in strip form by the use of mills or extruders, spray with anti-tack, cool and stack on pallets for fork-lift truck transport,

(b)   to shape the material into pellet or granular form by the use of extruders or pelletizers or by the use of mills and granulators, then to spray with anti-tack, cool and store in special storage for transport by conveyor or in bins by fork lift truck, and

(c)   to keep the material warm and feed it directly to calenders, extruders or presses. (This is by far the most cost-effective method if it can be arranged.)

5.1. Use of Dump Extruders

A post-mixer dump extruder has a large feed hopper which is often equipped with a pusher to force the material into the screw. The mixer discharges directly into this feed hopper. The mix is then taken into the screw and extruded down the screw flights to the discharge end of the extruder (head end).

There are many options available.

(i)   Slabber Heads (Fig. 3). These extrude the stock into a tube which is slit to form a sheet. This puts energy into the stock and usually the temperature of the compound rises as it passes through the extruder and head.

(ii)   Slabber/strainer head (Fig. 4). This has the advantage of filtering the mixed material to remove unwanted contamination, but the use of a straining filter increases the work done on the mixed material and so increases the temperature. A rule of thumb is that the output is reduced in proportion to the

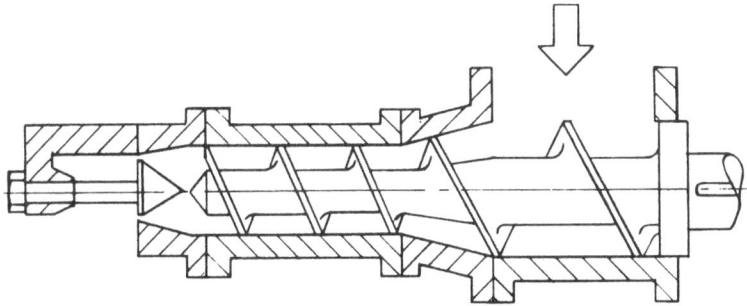

FIG. 3. Section through dump extruder with slabber head. © Farrel Bridge Ltd, 1987.

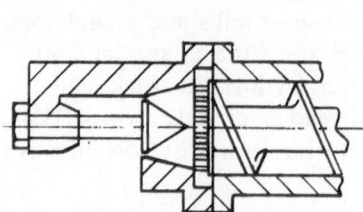 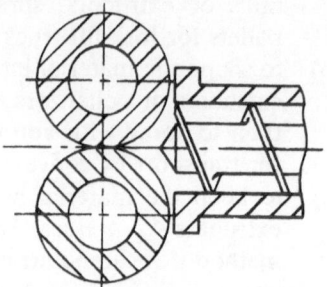

FIG. 4. Strainer slabber. © Farrel   FIG. 5. Two-roll sheeter. © Farrel
       Bridge Ltd, 1987.   Bridge Ltd, 1987.

size of mesh of the filter, i.e. 20s mesh by 20%, 40s mesh by
40% for the same output temperature. This rule of thumb of
course does not hold good for 100s mesh but is is useful to give
some indication of the effects of fitting a filter mesh. The
output falls as the pressure rises because the filter becomes
blocked with the strained material.

(iii)   Two-roll sheeter heads (Fig. 5) and roller die sheeter heads.
        These designs have the advantage of reducing or, at the worst,
        not increasing the temperature of the mixed material.

(iv)   Roller die head (Fig. 6) with contoured head and precision
        two-roll calender to make accurate sheet.

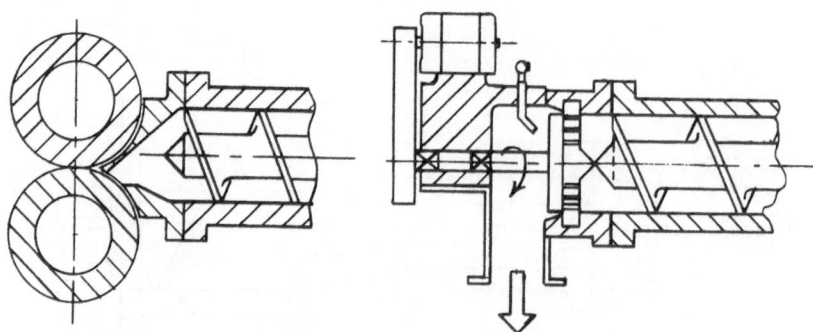

FIG. 6. Precision roller die. ©   FIG. 7. Pelletizer. © Farrel Bridge
       Farrel Bridge Ltd, 1987.   Ltd, 1987.

(v) Pelletizer head (Fig. 7) will with no, or at most a small, increase in temperature of the mixed material shape and cut it into cylindrical pellets. Pellets are mostly 20 mm diameter and vary in length from 10 to 30 mm. This length variation reduces during a production run as the machine temperature stabilizes.

The advantage of pellets is the big surface area:volume ratio which enables rapid cooling to take place and the ability to blend several batches together to improve product uniformity. To cut the pellets it is necessary to lubricate and cool the knives and also to use an anti-tack material to prevent the pellets sticking together. Ideally the temperature of the lubricating and anti-tack solution should be high enough for the heat of the pellets to dry off the surface moisture leaving a coating of the anti-tack material on the pellets.

Lubricating and anti-tack solutions are water-based and include a wetting agent (soap or detergent) and a powder such as magnesium silicate, calcium carbonate, etc.

Storage of pellets must be carefully designed so that they are not subjected to high compressive loads or kept in motion too much so that the anti-tack is worn off them. The pellet storage bins should empty completely for easy formula change. Whilst the pellets are ideally kept at 15–20°C, very low temperatures of the pellets can lead to processing difficulties, resulting in extended mixing times, high loads on the extruders and lumps in the stock.

5.2. Use of a Dump Extruder Instead of a Two-Roll Mill

A dump extruder should be used

(i)   when it is not possible to mill the compound because it is too sticky or too friable,

(ii)   when it is required to filter the mix material. If the mixer cycle is over 4 min it is better to cool the batch on a mill before feeding it to the strainer extruder,

(iii) when the mixer is on very short cycles; it is often the ability to mill the compound which restricts the production rate. At over 20 mixing cycles per hour, a dump extruder with a two-roll sheet head or a roller die should be considered,

(iv)   for automated plants with minimum manual labour.

6. TWO-ROLL MILLS

6.1. Use of Two-Roll Mills

By milling the mixed material it is possible to

(i)   reduce the temperature of the material,
(ii) make the stock more uniform both in distribution of the mix
      and the temperature of the mix,
(iii) do more work on the material to reduce the viscosity, and
(iv) disperse sulphur/accelerator.

Generally is is not possible to improve greatly the dispersion of carbon black, etc., or to reduce hard agglomerates by additional milling.

6.2. Milling Problems

Which roll will the stock stick to—front or back? Most mills are arranged with a speed differential (friction ratio) between front and back rolls so one roll runs faster than the other. In general, natural rubbers and SBRs stick to the slower roll and EPDM, neoprene, etc., stick to the faster roll, but there are exceptions.

A change of temperature can help to keep the stock on one particular roll. Often the front roll is made into the faster one by a change of gearing, but in a mill room which has a wide range of stocks it is always a problem. The use of mills with variable friction ratio, with hydraulic drives to achieve this, are being specified at this time.

One solution which works is to change the gearing so the rolls run at even speed, i.e. the same speed on front and rear rolls. The mill motor will have an increased load on it by this change due to the greater aggregate of the roll speeds, but the operator can now have some control over the roll to which the stock sticks.

6.3. Number and Size of Mills to be Used with a Batch Mixer

One mill when used with an internal mixer is normally sufficient to deal with up to 15 batches per hour.

For tyre masterbatches single mill units can process up to 30 batches per hour feeding direct into a strip cooler.

The normal surface roll speeds of mills are:

Size 1500 mm mill   24·4–27·5 m/min (80–90 ft/min)
Size 2200 mm mill   27·5–33·5 m/min (90–110 ft/min)

A batch from a 70-litre internal mixer when sheeted out to 600 mm wide × 10 mm thick makes a sheet 9·3 m long (i.e. it can be stripped off the 1500 mm mill into a strip cooler in 21 s) and a batch from a 270-litre internal mixer can be sheeted off a 2200 mm mill into a sheet 800 mm × 10 mm thick and 27 m long in 53 s.

However, the use of additional mills can increase the productivity and flexibility of a plant. The increase in milling will reduce the viscosity of the compound and reduce its temperature and improve the surface finish of the sheet. Another advantage of additional mills is to increase the 'holding' capacity of mixed stock should there be a hold-up in the discharge end of the line, e.g. cooler jammed, shortage of pellets.

With a two-mill set-up, the first mill will roll out the batch and quickly reduce the temperature. It is usual to have an operator on this mill to do several cuts to allow the stock to pass through the nip and then send the stock on a conveyor to the second mill. This mill can be fitted with a stock blender and a variable speed drive so that the stock can be continuously fed into the rubber strip cooler at a constant speed. For example with a 270-litre internal mixer on a 3-min mixing cycle, the speed of a 800 mm × 10 mm strip would be 9 m/min. This continuous lower speed allows for better cooling and drying and avoids too many splices in the strip between batches.

7. CONTINUOUS MIXING

The continuous mixing of rubber compounds is possible if the components of the mix are all in granular, powder or liquid form.

7.1. Rubber Granulation

Bales of rubber must be comminuted into granular form if they are to be used in a continuous process. When a bale is cut into granules it is necessary to use an anti-tack powder or liquid to stop the material sticking, the amount of anti-tack depending on the size of the rubber particle. A material consisting of 1 mm cubic particles has ten times the surface area of a material consisting of 10 mm cubic pellets (e.g. 1 mm ≡ 5·3 m²/kg, 10 mm ≡ 0·53 m²/kg). Therefore 1 mm material requries ten times the anti-tack of 10 mm material.

Granules of 3 mm cubes have a surface area of 1·78 m²/kg and it has been reported that 5% of anti-tack is required. Therefore 1 mm cubes need 15% and 10 mm cubes require 1·5% anti-tack. This anti-tack can contaminate a mix and therefore should be chosen with care.

Many synthetic rubbers are already available in pellet or granular form.

7.2. Machine Utilization
The concept of continuous mixing is based on the idea of mixing rubbers, powders, oils, etc., in a continuously fed and discharged machine which has a steady electrical load and is operating at a steady temperature, so the heat transfer capability of the machine is used efficiently. A batch mixer is empty and unloaded 10–30% of the time and the heat transfer is only used 50% of the time.

Continuous mixing machinery is utilized 100% of the time during a production run, and plants are in operation which run through meal breaks and shift changes, so operator involvement is not required all the time. With batch mixing the continual presence of labour is necessary in by far the greatest number of cases.

The ideal way to mix continuously would be to have continuous weigh feeders which would continuously feed in the correct ratio to give the desired formula of rubber, chemicals and liquid directly into the continuous rubber mixing machine. However, most rubber formulas have on average 15 different chemicals to make the complete formula, so it is very difficult to do this weighing continuously due to the high cost and varied nature of the chemicals involved.

7.3. Preblending
The method used today is to make a preblend of the chemicals in a powder preblending machine and use this amalgam to feed the continuous mixer.

Preblending uniformly distributes the ingredients and does the job of blending the mix (which in batch mixing is done by the helical shapes of the rotor, and cutting and rolling of the stock by the operator on a two-roll mill.)

Many compounds can be made into easily handled, free flowing premixes, especially those with mineral fillers (clay, limestone, etc.) and oils which make very stable preblends which can be stored, transported and continuously fed very effectively.

Other chemicals do not handle so well and the resulting preblend is unstable. This means that there is not a uniform distribution of the chemicals throughout the premix and the resulting mix when processed in a continuous mixer varies greatly. Chemicals which are unstable include large-particle rubber (*ca* 20 mm across), carbon blacks which the premixing operation breaks up into fine powder and talcs; generally, formulas which have little or no oils are unstable.

However, special storage bins are now available which can overcome this problem and the most difficult preblends can be dealt with without problems.

7.4. Premixers
The machines available fall into two categories.

(i) Machines with vertical rotors; these are of the type so successfully used with PVC preblends.

(ii) Machines with horizontal rotors and high-speed particle refiners.

Both types have been used with good results, but it is important that

(a) the machine empties each time it is used,

(b) the action of the rotor does not compact the chemicals and make agglomerates, and

(c) the rotor action does not aerate the premix.

Heating the chamber of the premix to about 50°C melts waxes and prevents a build-up of chemicals on the inside walls.

The storage of the premix compounds often requires special equipment but in all cases the storage should be 'live' with some form of rotor to break up blockages and bridges of the premix.

7.4.1. Method of Premixing
The general rules are as follows.

(i)   Feed polymers first. This breaks up any residue from the previous mix. This should be for a short time, as too long a time could result in the polymer sticking together.

(ii)   Add waxes, resins, sulphur and accelerators (if much wax is used the mixer should be preheated to prevent the wax sticking).

(iii) Add mineral fillers.

(iv)   Add blacks. Do not overmix blacks as the bulk density can reduce and aeration occurs.

(v)   Add oils. The addition of oils, which should be over as long a period as possible, stabilizes the mix and tends to increase the bulk density.

The speed of the premixer rotors should not be so high as to splatter the powder with great force onto the walls of the premixer and cause compaction and agglomeration.

High-speed particle refiners should have sharp edges on their blades, as blunt edges do not break up agglomerates, but tend to create them.

The premixer cycle is in the order of 6–15 min, to feed, mix and discharge.

Batch sizes should be approximately 35% of the free volume for vertical premixers and 60% of the volume for horizontal premixers.

7.5. Continuous Mixing Machines

It is possible to make a special extruding screw to mix and extrude rubber compounds continuously. A problem is that ideally the design of the screw, both in shape and length, should be changed to suit each compound to be processed, as different compounds require different work inputs and processing temperatures.

The author's design of continuous mixer, the Farrel Bridge Mixing and Venting Extruder—MVX—has two sections, one for mixing (see Fig. 8) and the other for extrusion.

The machine operates just as a heat exchanger, the power of the motors being converted into heat which all goes into the mixed compound increasing its temperature; the compound in turn heats the

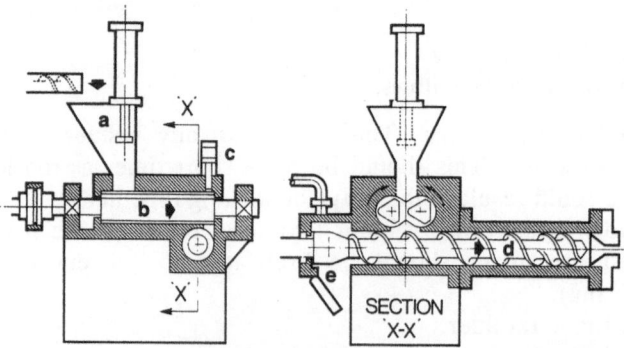

FIG. 8. MVX basic configuration. Key: a, inlet hopper with pusher; b, mixing chamber; c, vent port; d, extruder screw; e, reverse port to empty screw. © Farrel Bridge Ltd, 1987.

metal so that heat energy can flow from the rubber to the cooling water.

7.6. Computer Control

The operation of continuous mixing machines is boring. The first hour or so is interesting, but after that the premix just goes in and the compound just flows out. The excitement of feeding a batch mixer every three minutes and pushing buttons and watching temperatures, times, etc., is just not there and an operator's mind soon wanders away from the task.

This is where the computer comes into its own; it can be programmed to watch continuously and mind the machinery 100% of the time.

The steady-state running of a continuous mixer is ideal for 'feed back' control loops, which can be programmed for example to control the speed of the mixing motors from a thermocouple signal, or to change the temperature of the machine components as the mixer accelerates in production from start-up to running conditions.

The program can monitor the lubricating, the electrical equipment, air and water services, and take the necessary steps should a failure occur. The installation, commissioning and operation of computer-controlled equipment takes longer than that of normal electromechanical equipment. The electronic equipment of the microprocessor has a high fault level at the commissioning stage (six faults on one recent job and eight on another, both with equipment from different suppliers) but once the initial problems are overcome the microprocessor, with its associated analogue-to-digital converters, is very reliable.

The temptation to 'short out' some of the many flow switches, level switches, pressure sensors, etc., must be resisted and they must all be fully operable. Equipment relies on these signals for its protection in the absence of operators who can use sound, sight and smell to detect problems. Often they have been 'shorted out' to overcome some modification at the commissioning stage, so that considerable damage to equipment has taken place later when these signals are really required to protect the machinery.

7.7. Direct Product Production

The most efficient method of manufacture is to do the mixing and shaping to final product in one heating operation. Polymers which can be mixed in one stage are suitable for this method.

If it is necessary to masticate and break down the polymer at high temperature to obtain the required viscosity, then it is not possible to mix directly and extrude the product. However, with treads it is possible to use a continuous production method to do the final mixing of masterbatches and the shaping into a tread section.

The MVX has been used to make hoses with a cross-head, and sheet material with a roller die head. This method is economical when the whole of the output of the MVX can be used. If, for example, small extruder lines with an output of 50–200 kg/h are to be used, then direct compounding would be difficult to justify. But for 400–3 000 kg/h direct compound and product production has proved to be very profitable.

Chapter 7

QUALITY REQUIREMENTS AND RUBBER MIXING

P. S. JOHNSON

Polysar Ltd, Sarnia, Ontario, Canada

1. INTRODUCTION

To the casual observer, there might not appear to have been any significant developments in rubber mixing in the last decade. It still has all the appearances of a black art, in both the literal and the metaphorical sense. However, the oil crisis of the early 1970s, and the subsequent inflation in the cost of petroleum hydrocarbons and of energy, marked a turning point in rubber processing generally, and in mixing in particular.

A chapter on rubber mixing written about ten years ago, for example the excellent review by Palmgren,[1] would have dealt with the theory of dispersive mixing, with mixer design, and with techniques. It would have said little about the need to increase the efficiency of energy utilization in mixing; nor would it have dealt with microprocessor control, or the need for increased consistency of the product of mixing. However, Palmgren's prescience was remarkable. The final section of his review, reproduced below, was headed 'Needs for Future Work'.

'For development of the rubber mixing process and increasing its efficiency and economics, much research work is needed in the following areas:

Development of theories for dispersive mixing which will elucidate the process and which can be used for simple quantitative calculations.

221

Development of methods which in a simple and exact way can judge the result of a mixing process to allow determination of either the "goodness of dispersion" or the time required to reach an acceptable level of dispersion.

Better publication of details of mixer design, especially that of the rotor, to increase the understanding of the flow behaviour in the mixer.

Improved analysis for hydrodynamic–rheological conditions in an internal mixer and proving necessary simplifications by suitably designed experiments.

Trials to reduce mixing energy requirements by modifications of raw materials used. The problem is concentrated on agglomerate strength and surface nature of pigments and powders to be dispersed. Improvements would have a tremendous influence on productivity in the mixing process, and put rubber manufacturing in a better position as energy becomes scarce.

Refined heat–energy balance and heat-transfer calculations in the mixing cycle.'

It is hoped that this review will show how well these needs have been answered in the ten years or more since Palmgren's review was written. There are three other general reviews, published in the intervening years, to which the reader is referred for further details.[2-4] References to specific subjects are also given in the review.

2. THE MIXING PROCESS

Until recently, it has been usual in the rubber industry to consider all aspects of processing, from raw polymers to finished product, as more of an art than a science. As an art, it had to be transmitted, on the factory floor, from master to pupil, by a sort of apprenticeship system. This attitude is slowly changing. The veil of magic, or art, is gradually being drawn aside, and some light is being shed on rubber processing, especially mixing, by the application of science.

In a compounding plant there is usually more concern about the properties of the end-product than about the process of mixing as such. However, the increased costs of raw materials and of energy have forced the industry to be more concerned about the efficiency of mixing. To the compounder or production engineer, developments in the understanding of the fundamental behaviour of elastomers are only of importance if they can be directly and simply applied to improving industrial-scale rubber processing. However, in the long term, a clearer understanding of the physical processes occurring in the mixer will enable better control and, therefore, optimization of the process.

2.1. Types of Mixing

When mixing fillers, plasticizers and chemicals into rubber, the aim is to produce a product that has the ingredients incorporated and dispersed sufficiently thoroughly for it to shape easily, cure efficiently, and give the necessary end-use properties—all with the minimum expenditure of machine-time and energy.

If one looks at mixing as a unit process, once the compound formulation has been decided upon, converting the separate ingredients into the mixed compound involves problems of rheology, heat exchange, engineering, and economics. Consideration has to be given to the effect of machine variables on the microscopic properties of the material, and, in turn, how these affect the properties of the end-product (Fig. 1).

The blending of raw rubbers and the mixing of fillers into rubbers are important phases of rubber processing, and it is surprising that, until recently, little work had been done on the mixing process. The terminology used is often vague, overlapping, and confusing. Hindmarch and Gale's[5] concise definition of the mixing process, and of the two important but distinct actions, helps to clarify the subject:

'Distributive Mixing: an operation which is employed to increase the randomness of the spatial distribution of the minor constituent within the major base with no further change in size of that minor constituent, e.g. distributing black masterbatch or accelerator masterbatch throughout a compound.

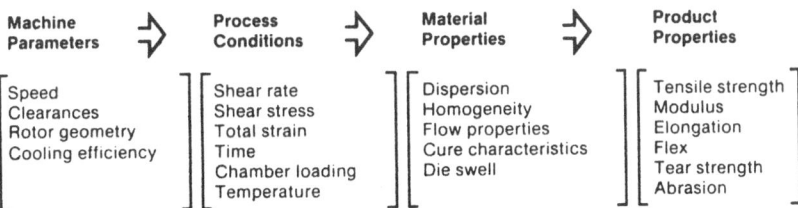

FIG. 1. The mixing process.

Dispersive Mixing: an operation which reduces the agglomerate size of the minor constituent to its ultimate particle size, e.g. compounding carbon black into polymer in an internal mixer.

This clearly is an over simplification when applied to practical mixing problems since both actions will occur simultaneously. Homogeneity will have been achieved when both the dispersive element and the distributive element of the mixing has achieved a fluctuation in average composition below a certain fixed and acceptable level.'

2.2. Incorporation

However, before either distributive mixing or dispersive mixing can effectively take place, there is an initial step in which the originally separate ingredients form a coherent mass. This process is known as incorporation or wetting.

Tokita and Pliskin[6] pointed out that in most processing operations rubbers do not deform smoothly or flow steadily, but rather yield and rupture. They also demonstrated that basic rheological properties cannot directly explain the reaction to applied stress of gum rubbers—the characteristics described in the industry as 'cheesy', 'nervy', or 'rubbery'. They developed a deformation diagram (see Fig. 2) which related deformation and extension at break to mixing behaviour, and showed that wetting or incorporation time was also related to elongation at break.

Nakajima[7–10] has shown that behaviour of a viscoelastic material in

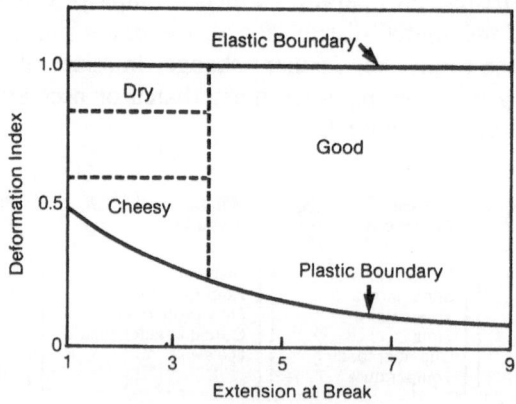

FIG. 2. Deformation diagram.

an internal mixer is more appropriately described as deformation rather than as flow. The mixing process imposes a large deformation, which often exceeds the breaking strain, and so material failure is important in mixing, especially in the incorporation stage.

Nakajima[7] showed that the incorporation step has two mechanisms. In the first, the elastomer undergoes a large deformation, increasing the surface area for accepting filler agglomerates, and then sealing them inside. In the second mechanism, the elastomer breaks down into small pieces, mixes with the filler agglomerates, and once again seals them inside. The former mechanism is easily observed in an open mill whereas the latter is not necessarily visually observable because the breaking and sealing steps occur on a micro-scale (Fig. 3). In an internal mixer, the material is forced through a narrow passage between the rotor blade and the chamber wall and, in the process, undergoes extensional deformation. At the same time, the differential velocity between the rotor tip and the stationary chamber wall also creates shear deformation. Cotten[11] believes that crumbing and tearing in the internal mixer can be caused by failure of the rubber in the extensional flow region, immediately in front of the rotor. Freakley and Wan Idris[12] consider that, with a material having low elongation at break, especially if the mixer is underloaded, such crumbing or fracture occurs in the void formed behind the rotor. Thus, any model of the mixing process has to take account of both extensional and

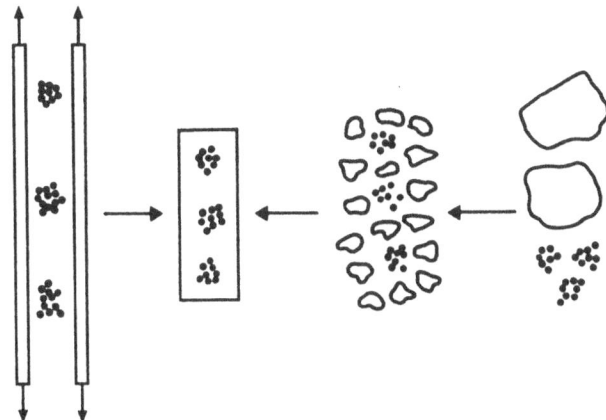

FIG. 3. Nakajima's two mechanisms of incorporation.

shear flows. Cotten[11] states that extensional flow is more efficient in incorporation and breaking-up rigid aggregates (dispersive mixing) than the shear flow at the same deformation rate, but because rubber mixers are designed to provide a region of very high shear, in practice extensional flow accounts for only a minor amount of the total energy dissipation.

Nakajima[9] also showed in an actual case study that the comminution mechanism, described above, accounted for about one-third of the energy of mixing.

In addition to the two mechanisms of comminution and stretching, elastomer and carbon black have to be compacted.[13] Compaction, or massing, is the displacement of entrapped air in the machine by applied compressive force. This requires deformation of elastomer domains to match the shape of the carbon black, followed by relaxation of the elastomer in the deformed state. In another case study, Nakajima and Harrell[13] estimated that about 6% of the energy used in incorporation was required for compaction.

2.3. Dispersion

In the early stages of mixing the carbon black forms relatively large (10–100 μm) agglomerates. During dispersion they are broken down to a size of less than 1 μm. The dispersion stage of mixing requires higher shear stress and energy input than the incorporation stage, and the physical properties of the mix alter as the dispersion proceeds.[14–16] The changes in power consumption in a typical rubber mix are indicative of stages in the process such as incorporation, dispersion, and plasticization, and can be related to the development of end-product properties (see Fig. 4). At the end of the incorporation stage, the majority of filler is present as rubber-filled pellet fragments. These act as large filler particles, whose effective volume is higher than that of the filler alone, both because of the rubber inside the particles and the rubber immobilized between them. As dispersion proceeds and these agglomerates are broken down, the rubber immobilized between the particles is released into the matrix.

Properties such as viscosity and die swell, which are related to the effective volume fraction of rubber in the mix, approach a steady-state value at the end of the dispersion stage. A well dispersed mixture always has a lower viscosity and a higher die swell than a comparable less homogeneous mixture.[17–19]

Attempts have been made to model the process of dispersive

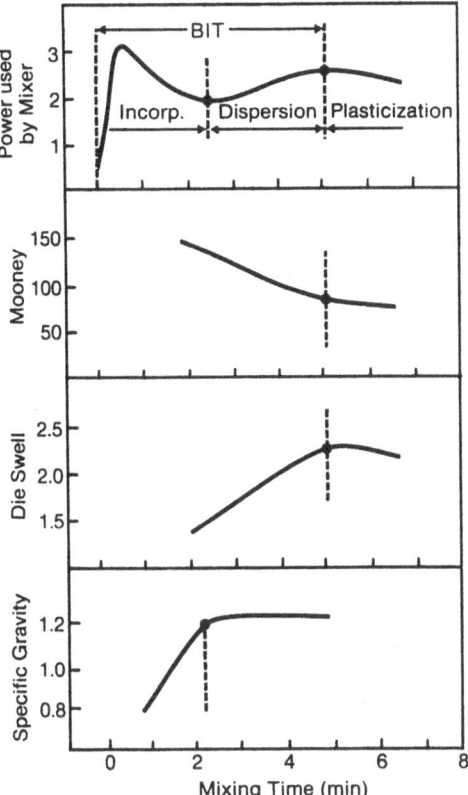

FIG. 4. Power usage and physical property changes during mixing cycle.

mixing.[20–26] The best exposition is the review by Tadmor and co-workers,[24] but Cotten's summary[23a] is clear and concise and is quoted here by permission of the author.

'McKelvey[20] considered a system of two particle agglomerates suspended in a viscous liquid under shear. His equation predicts that some minimum force must be exceeded before agglomerates will break. Additionally, only agglomerates oriented properly to the direction of forces will be broken; thus, the flow pattern must be such that the orientation of agglomerates is continuously changing. The ease of dispersion is proportional to the size of individual particles.

The previous analysis was recently extended by Tadmor[21] to agglomerates formed by dissimilar particles in shear and elongational flow fields. In a subsequent paper,[22], Zloczower, Nir, and Tadmor attempted to fit their model to the actual mixing of rubber compounds in an internal mixer, assuming that dispersive mixing is dominated by the agglomerate rupture in a narrow gap, high shear field. Tadmor's analysis predicts that the rate of agglomerate rupture is independent of agglomerate size, but rather depends on the size of aggregates.'

Cotten's experimental study[23] of the mixing process, using power curves for Brabender mixes, is an admirable refinement of the technique originated by van Buskirk and co-workers.[27] His finding, that the rate of dispersive mixing increases with decreasing surface area and increasing structure of aggregates, is in agreement with Tadmor's prediction that the rate of agglomerate rupture depends on the number of particle–particle contacts and, thus, is related to the size of individual aggregates, but is independent of agglomerate size.

Another excellent experimental study, using a highly instrumented laboratory-size Banbury and biconical rotor rheometer of mixed batches, is described by Freakley and Patel.[28] They performed a detailed analysis of flow and mixing characteristics in the region of a rotor wing. One interesting finding is that dispersive mixing, which depends on the stress levels generated during mixing, is shown to occur throughout the entire mass of material swept in front of the rotor wing and not simply at the rotor tip. Furthermore, the stress levels depend more strongly on batch temperature than on rotor speed.

This work is described, in the introduction, as the first stage in the formulation of viable mathematical models by adopting a practical approach to the problem of mixing. The present author knows, from personal communication with Professor Freakley, that it is equally the fruit of many previous studies. Future papers in this series will be eagerly awaited.

2.4. Process Variables

2.4.1. Heat Exchange
A major concern in most mixing operations is to control the temperature at the end of the mixing cycle, and so it has been normal to use cold or chilled water as a coolant. With the advent of internal

mixers having a more efficient heat exchange capability, it was found that cold water can be too efficient a coolant at the beginning of the mix, the rubber slipping on the cold metal surfaces. This can be prevented by maintaining the cooling water at such a temperature that the rubber adheres to the chamber wall and rotors and deforms readily.[29] This results in more consistent mixing and a slight reduction in mixing time. Nakajima[13] reports that this is seen as a significant reduction in the time required for ram seating, i.e. compaction.

A major problem in cooling rubber, in any process, is its inherently poor conductivity, which means that it is only cooled where it touches a cool surface. Thus, total heat removal depends on the area of the cooling surface and also on the way in which fresh rubber surfaces are moved into contact with that cooling surface. In internal mixers this depends mainly on the geometry of the rotors. Mixers with intermeshing rotors are less influenced by the friction or adhesion between the rubber and metal, and are, therefore, less sensitive to starting temperature.[30] Rotor geometry also affects the overall heat transfer coefficient in a mixer.[31–35]

2.4.2. Dump Criteria

Time and temperature have been the most commonly used criteria to determine when to terminate the mixing process. The aims are to guarantee the quality of the end-product, avoid overmixing, and reduce variation between batches. There is an increasing body of evidence that a more precise and reproducible control of the mixing cycle can be obtained by following the energy input at various stages in the cycle. This is because the changes in power consumption are indicative of the wetting, dispersion, and plasticization stages in the process, as described earlier.

Mixing to a preset time does not allow for variations in metal temperature at the start of the mix, for cooling rate, or for ingredient addition times. This can result in significant batch-to-batch variation. When mixing to a predetermined temperature, as is often done with upside-down mixes having short mixing cycles, the major limitation is the accuracy with which the batch temperature can be measured. The large heat-sink provided by the machine often makes temperature measurement inaccurate, though there are now infra-red probe thermocouples available that are more accurate.

Mixing to a predetermined power input into the batch overcomes these limitations and gives improved batch-to-batch consistency with mixes requiring longer than a 3-min mixing time. However, following the work or energy input alone is not sufficient. In addition, the effect of process variables on the shape of the power curve has to be established. In other words, a recording chart, which indicates both the instantaneous power and the integrated power or work done, is required. Furthermore, these data should be considered additional to the established control criteria of time and temperature and not as a completely separate alternative set of criteria.[18,19,27,36-39]

2.5. Operating Variables

Although the influence that changes in operating variables has on the subsequent processing behaviour of rubber compounds has been recognized for a long time, the separate effects of those changes are difficult to determine. This is mainly because there are strong interactions between the variables.[40,41] However, there are some general points that can be made.

2.5.1. Rotor Speed

This directly affects total shear strain or deformation and thus the speed of mixing.[3] The speed of mixing is usually limited by the maximum allowable temperature, i.e. by the balance between heat generation and heat removal.[1,2] Dispersive mixing, although dependent on shear stress, does not seem to be directly affected by rotor speed; this is probably due to the effect of elongational flow, which also creates high shear stress.

However, the reduction in viscosity which results from temperature rise, results in decreased shear stress and, therefore, decreased dispersive action. Thus, there is a trade-off between increased speed of mixing and less well dispersed or homogenized product. Therefore, in most commercial mixers there is a limit to practicable rotor speeds.

2.5.2. Ram Pressure

The main function of the ram is to keep the ingredients in the mixing area. In addition, high ram pressure has definite advantages, especially for high-viscosity mixes.[1-3] High ram pressures decrease voids within the mixture and increase shear stress by reducing slippage. This is because the effect of increasing pressure is to increase the contact force between the rubber and the rotor surface, thus altering the critical stress so that flow begins at a lower temperature.

2.5.3. Chamber Loading

It is important to optimize chamber loading.[12,31–35,37] Sufficient material is required in the chamber to produce the ram pressure effects described above. If the mixer chamber is excessively overloaded, much longer times are required for good dispersion and there is the danger that there will be undispersed polymer. With severe underloading, poor mixing may also result. The flow patterns in internal mixers are poorly understood, but it appears that the formation of voids behind the rotor tips, and the resulting turbulent flow, are necessary for good mixing.[12] The optimum batch weight for a particular mix depends on the type and level of rubber, filler and plasticizer. Arbitrary use of 'fill factors', or other rules of thumb, is not recommended.

2.5.4. Order of Ingredient Addition

This is the area in which the compounder's expertise is the most evident and most needed. The number of alternatives is large and the optimum order of ingredient addition depends on the size, type and degree of wear of the machine, the available speeds and ram pressures, and the types and levels of rubber, filler, plasticizers and minor ingredients.

There are a number of papers in the literature that are useful,[3,4,37,42,43] and although each mix has to be considered separately, there are some general rules. For example, it is generally preferable to add fillers early in the mixing cycle so that good dispersion is achieved because of the higher viscosity, and thus higher shear stress, at the lower temperature then prevailing. For the same reason, oils and plasticizers, which reduce viscosity if present in large quantities, can slow down dispersion, and are usually added later.

However, if oils and plasticizers are added after the fillers are incorporated, they can coat the metal surface and, by acting as a lubricant, slow down both distributive and dispersive mixing. For this reason, upside-down mixing procedures, and techniques such as adding oil and absorbing filler together, are used.

2.5.5. Scale-Up

Valid scale-up criteria from laboratory-scale to industrial mixers are difficult to develop because of the inaccuracy of the models and hydrodynamic analysis.[1] Funt[2] lists the alternative base criteria as:

(1) simple geometric similarity,
(2) constant maximum shear stress,

(3) constant total shear strain,
(4) constant work input,
(5) constant mixing time,
(6) constant stock temperature history,
(7) constant Weissenberg or Deborah numbers,
(8) constant Graetz or Griffiths numbers.

Morrell[3] summarizes the arguments for and against each of these, both for mills and mixers. The review by Tanaka[44] is also a useful summary of the various approaches.

Tadmor and Manas-Zloczower[45] conclude that the two most successful criteria in practice have been the unit work input,[2,27,36] and the total shear strain concept,[2,27] especially when temperature profiles and volume loading for different machines are closely matched. They propose a new scale-up criterion, derived from their theoretical model of the dispersive mixing process mentioned earlier.[22]

3. MIXING EQUIPMENT

There have been three significant developments in mixing equipment in the last ten years: (a) an increased understanding of the influence of rotor design on mixing efficiency, (b) the application of automated microprocessor control of feeding, mixing and discharge from the mixer, and (c) development of continuous mixing equipment. Only the first two will be dealt with here. The third development, continuous mixers, is dealt with by Ellwood in Chapter 6 of this volume.

3.1. Rotor Design and Mixing Efficiency
Meissner and Reher[41] consider that the mixing process can be divided into three basic elements, which interact with one another. These are the mixer geometry, the properties of the material, and the operating variables. A number of companies make internal mixers for the rubber industry, and whatever their design, they have two common features. These are: (a) the ability to exert a high, localized shear stress on the material (a nip-action), and (b) a lower shear rate stirring or homogenizing action. It is the combination of these two effects, high shear stress plus large shear deformation, that is effective in achieving both dispersive mixing and distributive mixing.

To some extent mixer rotors can be designed to meet specialized requirements, but the majority are of a compromise design, based on experience, to fit best the wide range of compounds typical of a mixed production.[3]

There are two basic designs of rotor: tangential and intermeshing (see Fig. 5). Tangential rotors, like the rolls of a mill, can rotate independently and, therefore, at different speeds. Intermeshing rotors, like gear wheels, if of the same diameter, have to rotate at the same speed. Both types have had their proponents, but until recently there was little factual evidence in the literature to enable an objective comparison.

Ellwood[29] showed that in the Banbury mixer, which has tangential rotors, it is necessary to adjust the temperature of the metal in contact with the rubber in order to achieve the necessary degree of adhesion between them.

Whitaker showed[30] that this did not apply to the Shaw Intermix, which has intermeshing rotors; he claimed that because of the more positive mixing action with this design of rotor there is, in fact, no problem with slipping at the start of mixing.

Wiedmann and co-workers[31] studied heat generation and heat transfer in tangential internal mixers and showed that the important factor is contact surface between the compound and the wall. They

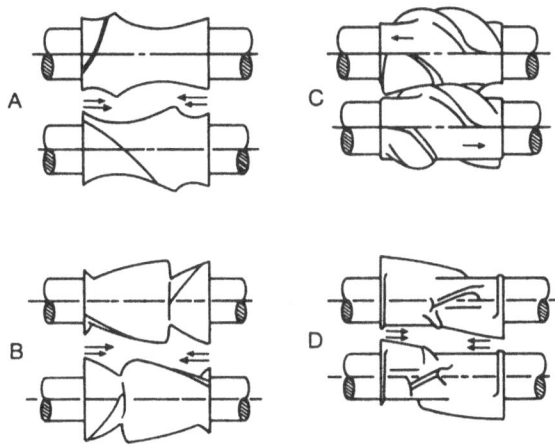

FIG. 5. Examples of rotor designs: A, Banbury two-wing; B, Banbury four-wing; C, Shaw Intermix three-wing; D, Werner and Pfleiderer four-wing. The pumping action of the wings is indicated by arrows.

extended this work[32-35] with the aim of optimizing the design of the tangential rotor and, also, of producing an optimun intermeshing design.

Their approach was based on the concept that throughput, or production, is dependent on batch size, and inversely proportional to cycle time and the number of mixing stages required. Batch size, for a given machine size, obviously depends on small rotor volume and high fill factor. Cycle time and number of stages depend on rotor geometry and operating variables, especially rotor speed.

Two important points are that intermeshing rotors have larger volume than tangential rotors with the same centre distance, and in general they operate at lower fill factors. Therefore to achieve the same batch size, a chamber larger by a factor of 1·65 is required for intermeshing rotors. As capital cost is roughly proportional to chamber size, this is a significant disadvantage for intermeshing rotor machines.

Wiedmann and co-workers found, as did Basir and Freakley,[40] that there are strong interactions between the variables, but they identified fill factor and rotor speed as the two largest influences on mix quality and dump temperature for each geometry. The next most significant variable was found to be ram pressure. They showed that temperature control of the mix was better with intermeshing geometries. This is because the action of the intermeshing rotors forces the rubber through the nip and also turns it and renews the surface with little shearing. At the same time the wings set up turbulent flow patterns which result in more rapid thermal homogenization of the material. This agrees with Whitaker's conclusions.[30] Wiedmann and Schmid have shown that the mixing action of intermeshing rotors achieves a better overall heat transfer, and that this accounts for the better temperature control.

In summary, this work showed that conventional rotor geometries, both tangential and intermeshing, can be improved in terms of greater working volume, quicker incorporation, better mixing, and better temperature control.

Optimized designs of both the tangential and intermeshing rotor machines are now available. The choice of rotor design depends on the type of compound to be mixed. The tangential system gives high machine efficiency, i.e. fast feeding, fast discharge and fast incorporation, and is thus better for short mixing cycles and multi-stage mixing. The intermeshing system is better for compounds that are difficult to disperse or give more rapid heat generation. With such compounds a higher production rate in kg/h can be achieved because fewer mixing stages are required, even though the mixing cycle is longer with the intermeshing geometry rotors.

3.2. Process Control and Mixing

The advances and cost reductions that have been made in electronics and, especially, in microprocessors, are, at last, being applied in rubber mixing. However, microprocessors are used much more in industry in general than in the rubber industry. Howgate and Cottey[46] consider that there are four reasons for this:

(a) A lack of quantitative understanding of the interrelationship between processing parameters and product quality.
(b) The high degree of dexterity, skill, or judgement required in many processes making automation difficult and expensive.
(c) A reluctance to invest in products that seem to be outdated almost as soon as they become available.
(d) Difficulty in interpreting and developing the typical claims and jargon of microprocessor equipment manufacturers.

However, the degree of consistency which is now being required of the product of mixing can only be consistently met if the mixing process itself is controlled and consistent. It is difficult to see how this can be achieved other than by microprocessor control. There are a number of systems available[35,47–49] that provide automated handling, weighing, and charging of all raw materials to the mixer and control the mixing process with a microprocessor. This monitors and/or controls temperature, energy input, number of rotations of the rotors, and mixing time. The developments in this field were described in a series of papers in an ACS Rubber Division Symposium on *Applications of Microprocessors in the Rubber Industry* in 1983.[50]

Most of the systems that have been installed concentrate on the materials handling as much as on control of the mixing process itself. Early systems were controlled using punched cards which gave instructions on weighing and addition of raw materials and on dump criteria. Basically, these systems reduced the chance of human error. More recent systems do all of the above, but also include a programmable controller. One important advantage of these is that they can provide a printed record of each batch. Details of the capabilities of the commercially available systems can be obtained from the manufacturers.

Power profiles—recordings of the power consumed during mixing in an internal mixer—have been used for some time, and their importance to the developing science of rubber mixing has been recognized by a number of authors.[14–16,18,19,36–39,51] The use of power profiles by the compounder has become so widespread that several equipment suppliers have produced instruments which will monitor the power required for mixing and provide process control action. In fact, most 'off-the-shelf' programmable controllers can provide suitable control action.

However, all of these devices require that the compounder has a reasonable understanding of the relationship between the power consumed by the mixer and a desirable process control strategy. This can be a reasonable requirement if the compounder has to consider only a limited number of formulations, or if runs are significantly long for each of the mixes. The compounder may use the technique known as evolutionary operation, 'EVOP', to determine a control strategy, and an appropriate instrument to effect that control action.

To develop a microprocessor control system for an internal mixer, in-house, is not an easy task. Before taking that route, in preference to purchasing a system it is suggested that the reader considers the experiences detailed in ref. 52.

In spite of the problems, one confident prediction that can be made for the rubber industry is that the use of microprocessors in process control will grow significantly in the next few years, though its growth may be slower in the mixing area than in other processes.

4. PROCESSABILITY TESTING AND MIXING

Processability of elastomers, whether in mixing, extrusion, moulding or curing is, and always has been, a problem for the processor. The concept itself is difficult to define. In fact, rubber processability is a subjective concept in that it depends on what you want to do with the material, and in what equipment, whether it processes well or badly. Furthermore, there has been a lack of simple, easy tests that can measure processability. Generally, if the test is simple it does not correlate directly with processability, and, if it does correlate, it is not

simple. A third problem is that, even when almost complete information is available on the elastomer or compound, in terms of its
viscosity, elasticity, ultimate elongation etc., the relationship of these
parameters to processability in a mixer or extruder, or to flow in a
mould, is not clearly understood.

4.1. Requirement of Processability Testing

Processability, when applied to the mixing of rubbers, means the ease
with which the compounding ingredients can be incorporated and
dispersed in the rubber. However, there are two basically different
requirements of processability testing. The first is in compound,
process, and product development where very extensive testing may
be justifiable. The other is in process and quality control, where only
enough data are required to enable rapid decisions about process
parameters or product release. Much of the effort in recent years has
been to automate and speed up test techniques, in order to transpose
them from research or development tools to process or quality control
tools. The other thrust has been towards understanding and predicting
the effect of fundamental properties on processing behaviour.

Demands for productivity, less scrap, and fewer rejects are increasingly placing greater demands on the control of processing methods
and materials. Efficient processing depends very much on the uniformity of materials at each stage of production. The sometimes
anomalous behaviour of unvulcanized elastomers and compounds, as
they are processed in mixing and forming equipment, are probably the
major concern of the rubber industry today.

The basic problems of process control in the manufacture of rubber
products, and the need for means of assessing the processability of
rubbers, can be summarized by the following three points.

(a) There is a need for material characterization at a number of
     stages in the production cycle, especially (i) incoming material
     and (ii) mixed compound.
(b) The property measured must relate to a significant response of
     the material to processing conditions (compound), or to its
     molecular characteristics (raw rubber).
(c) The method of measurement must be consistent with the
     practical constraints of manufacturing operations. This means
     that the test procedure must be rapid and amenable to
     execution and interpretation by semi-skilled or non-technical

staff, though for development work more complex techniques can be considered.

In recent years, there have been a number of developments that have increased our understanding of the physical processes that take place during the mixing, milling, extrusion, and injection moulding of elastomers. At the same time, techniques and rapid-acting instruments have been developed for measuring various aspects of processability. The combination of these developments should lead to increased efficiency in materials, in energy, and in equipment utilization.

Processability of raw rubber in an internal mixer depends on polymer type, molecular weight distribution, degree and type of branching, physical form, catalyst and emulsifier residues, stabilizer and other additives. In addition, filler type, filler concentration, polymer–filler interaction, and the levels and type of oils and process aids affect the process.

Measurements can be made on the raw materials in order to determine how well they will mix, and on the product of mixing, to see how well it has, in fact, been mixed. A brief outline follows of testing in relation to mixing, further discussion of rheological principles, and the types of instruments available. Practical applications are given in the review by Norman and Johnson.[53]

4.2. Testing Raw Materials

At present, there are only very tenuous links between raw polymer properties and either processing or end-product properties. However, it can hardly be doubted that polymer consistency and quality have an overriding influence on both processing behaviour and end-product properties. The advances that are being made in our understanding of basic molecular parameters, rheological properties and elasticity, and also recent advances in electronics, microcomputers and instrumentation, will inevitably mean that such links will become firmer.

4.2.1. Molecular Structure and Rheological Properties

Mieras[54] reported work aimed at relating molecular structure of raw rubbers to the processing properties of their black-filled compounds by relating both to the rheological properties of the rubber. However, he noted two limitations; firstly, that there is no theoretical or experimental means to characterize a rubber fully in rheological terms, and secondly, that it is virtually impossible to simulate processing situations in simple rheological experiments. His overall conclusion is that, although molecular parameters can be related to measurable rheological properties, it is probable that they are less important in processability than such characteristics as the strength properties of the elastic liquid.

As was previously mentioned in Section 2.2 on incorporation, Tokita and Pliskin[6] showed that basic rheological properties and failure characteristics of elastomers could be related to molecular weight distribution and also to processing behaviour. They related the Mooney viscosity and Mooney recoil, measured at several rotor speeds, to molecular weight distribution. The results showed that the broader the molecular weight distribution, the more non-Newtonian is the flow behaviour (i.e. the stress is less sensitive to the shear rate). They concluded, like Mieras, that basic rheological and physical properties are useful for polymer specification, but do not help much in specifying better processing polymers. Their major contribution was to point out that the deformations which the elastomer experiences are so large and so rapid, it has no chance to respond as a viscous fluid in most initial processing operations, such as filler mixing and polymer blending in an internal mixer, or on a mill. Thus, it does not deform smoothly or flow steadily, but rather it yields and ruptures. They showed that the molecular weight distribution (MWD) directly affects failure characteristics, particularly ultimate elongation. This is because a broad-MWD polymer experiences less stress in deformations than does a narrow-MWD polymer due to its non-Newtonian characteristics. In addition, its tear propagation rate is slower than that of a narrow-MWD rubber because of its higher relaxation time. Relationships between deformation characteristics and extension to break have been used to explain the mixing and milling behaviour of different rubbers.[8,10,55] However, the method is probably neither sufficiently precise, nor selective enough, to be useful in predicting processability of different grades of one specific synthetic rubber.

4.3. The Mooney Test

The Mooney test, which is used routinely to assess the processability of both raw stock and compounds,[56] relates to average molecular weight (M_w). However, as the molecular weight distribution has a considerable effect on the processing behaviour of rubbers of the same M_w, it is sometimes inadequate.[57] The Mooney test makes no estimate of elastic properties, and it works at a shear rate much lower than most processing operations.

Despite these acknowledged limitations it is still regarded, by the industry as a whole, as a most useful processability tester. The value of the Mooney test for assessing raw rubbers is as a benchmark. It is the basis for current specifications and is the major parameter for empirical control of polymerization reactions.

There have been a number of developments of the Mooney test. The generally useful and accepted one has been the Delta Mooney test, which has achieved the status of an ASTM test, and is well recognized within the industry. Other techniques which have been reported are Mooney recoil,[6] initial Mooney torque peak (PMT),[58] and variable-speed Mooney tests.[8,59,60]

The Delta Mooney is defined as the difference between two Mooney viscosity readings, either at two fixed times or between the early trough (b) and the subsequent maximum (c), as shown in Fig. 6.

An interesting development is the TMS (Turner, Moore and Smith—Avon Rubber, Wiltshire, UK) Rheometer,[40,61,62] which is based on a Mooney machine, but the rubber is injected into the closed cavity by means of a transfer pot mounted directly above the cavity. By this means, rubber with freshly created surfaces can be made to fill the cavity and a control maintained on the pressure inside the cavity. This arrangement gives reproducible results at shear rates of up to

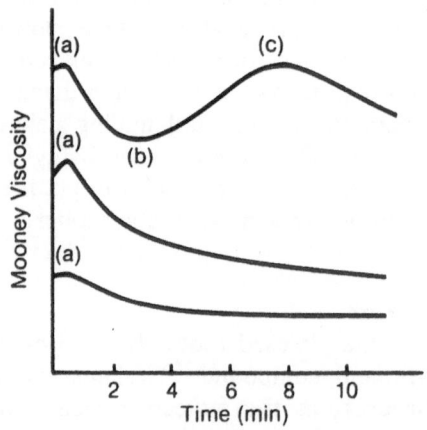

FIG. 6. Typical Mooney viscosity curves.

40 s^{-1}. Not only can shear stress for a given shear rate be measured, but stress relaxation and recovery can also be examined.

4.4. Capillary Rheometers

The majority of published work on extrusion behaviour deals with compounded stock. Those papers reporting work on raw rubbers[63,64] have usually been on the use of capillary rheometers to determine extrusion properties at higher shear rates than are possible with Mooney viscometers. Capillary rheometers are, in principle, quite simple to use, and the application of electronic, minicomputer and laser technology has reduced the operation and data analysis to a routine task. There are no standard ASTM or other test procedures, but under a specific set of conditions, once a material is characterized, the data can be used as standard for comparison of all subsequent batches. It is readily possible to characterize a raw rubber by an extrusion experiment to determine the viscosity/shear rate curve, extrudate swell, and stress relaxation.[65,66] Both Sezna[67,68] and Karg[69] have shown how the Monsanto Processability Tester (MPT), a modified, computerized extrusion rheometer, can be used in predicting mixing behaviour. The MPT (shown schematically in Fig. 7) is a most versatile instrument. It has a larger than conventional barrel for minimal pressure drop in the barrel, a pressure transducer at the entrance to the orifice, a microprocessor system, and a laser device for

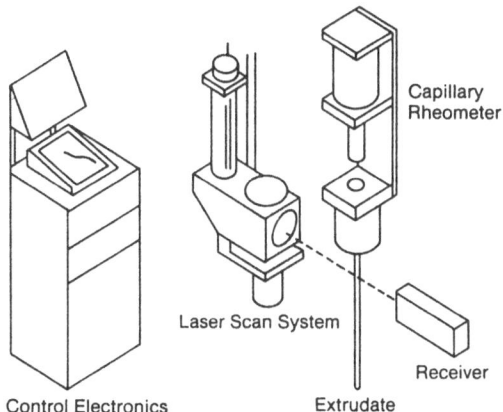

FIG. 7. Schematic drawing of Monsanto Processability Tester. Capillary rheometer with die swell detector.

measuring extrudate dimensions. Viscosity, die swell and stress relaxation measurements can be performed separately or in a programmable test sequence requiring less than 5 min. Scorch testing is also automated.

4.5. Stress Relaxation

The only stress relaxation tester commercially available is the Rapra/Monsanto SRPT (Stress Relaxation Processability Tester), although others have been described in the literature. The rationale behind this development is that viscosity alone is not able to predict rubber processability; some elasticity characteristics must be taken into consideration. Further, since most processing problems are related to memory or relaxation effects, it makes sense to consider using stress relaxation. However, once again, this technique has found most application with compounded stock.[70-73] It is claimed[71] that the technique is sensitive to small variations in molecular weight distribution. Experience so far indicates that it may not be sensitive enough to differentiate levels of processability within grades.

4.6. Miniature Mixers

The internal mixer is, in practice, the final judge of the processability (mixability) of a rubber although, obviously, one would hesitate to use a production-sized machine to test the processability of raw materials. There are several miniature internal mixers or torque rheometers available that can be used to assess the mixing behaviour of rubbers. It is not possible to separate the viscous and elastic responses of a material in torque rheometer traces, but they are useful for comparison between materials and for determining black incorporation time (BIT), which is defined as the time interval between the lowering of the ram and the second power peak (see Fig. 4). It should, perhaps, be called the 'black (incorporation plus dispersion) time', but the usage, as defined, has become hallowed by time. A laboratory mixer, equipped with a power recorder and integrator, is often used as a processability tester to assess the changes in resistance to deformation during mixing. The times to specific features of the power trace are used as indices of processability of the rubber, or of the dispersibility of carbon black in a given polymer.

There are three characteristics of a mixer power or torque rheometer trace that provide useful information.

(i)   The times at which peaks and troughs occur in the torque/time or power/time curves.

(ii) The total work done by the mixer on the rubber.

(iii) The early stages of mixing, or even the end of a first-stage mixing, may be indicated by the magnitude of the short-period variations in power or torque.

4.7. Testing The Product of Mixing

Most of the processability testing which is carried out in the rubber industry is performed on the product of mixing. There are three types of test: (a) tests aimed at determining the state of mixedness of the compound, (b) tests of some 'fundamental' property believed to be related to processing behaviour in subsequent operations, and (c) tests attempting to simulate those operations. Only tests of type (a) will be considered here.

4.7.1. Tests of Mixedness

Large inhomogeneities in mixed compounds are usually obvious to the naked eye and no special test instrument is required to detect them. However, the most important filler in the rubber industry is carbon black, which, by its opaque nature, prevents any fine assessment of homogeneity of mixing. In any case, elimination of highly loaded regions (homogenization) is not the only requirement. It is equally necessary to separate the tightly packed aggregates of black particles from each other to form a loose network (dispersion). It is the combination of these two processes which determines both the ultimate, or failure, properties and also the processing (viscoelastic) properties of the mix.[74]

Medalia, in his excellent review of carbon black dispersion,[75] divides the available techniques into three groups: (i) microscopy, (ii) surface roughness, and (iii) physical properties. Recently, there have been some interesting developments in techniques of the second type, measurements of surface roughness, usually of a freshly cut surface. Persson[76] describes a technique which allows simultaneous examination of the specimen with various controls using a split-field microscope. A recently described,[77] and patented, technique depends on determining surface roughness using a stylus pressed lightly onto the surface. Another interesting technique measures surface roughness by dark-field reflected light (DFRL) microscopy.[15] The image obtained from the DFRL microscope is relayed through a standard television system and displayed on a TV monitor, and also fed into an oscilloscope with a single-line strobing facility. That is, the instrument can scan any one line of a multiline television signal and display a trace of the intensity distribution across that line on its screen. The line manually selected for scan analysis by the oscilloscope appears as a bright pulse on the monitor. As many lines as required may be analysed separately by altering the location of the bright pulse and thus altering the scan region of the scope. It has been found that two scans per field are sufficient to characterize a sample. Yet other techniques which have been reported are small-angle X-ray scattering, dynamic mechanical measurements, scanning electron microscopy, and electrical resistance.[78]

5. QUALITY AND MIXING

The authoritative 1974 paper by Palmgren,[1] which has been used as the starting point for this review of developments in mixing in the subsequent decade, said little about the need for increased consistency of the product of the mixing process. Today, in retrospect, the most significant development in the rubber industry since 1980 has been the change in emphasis towards high quality and consistency of finished products. This change has been instigated by the end-product user, especially the automobile industry, and transmitted by the processor, back to the producers of the raw materials. This reflects a change in the philosophy of the final customer, who no longer willingly accepts the concepts of the 'throw-away' society or designed redundancy. It has been reinforced by competition, in price but especially in quality, from Japanese industry. Symposia on quality are becoming a regular item on the agenda for technical meetings[79,80] and many papers, both general[81-83] and specific to the rubber industry,[84] have been published.

5.1. Quality Insistence
The result of this change in emphasis has been that the development of more exacting specifications applied to rubber products has led to greater insistence on the quality and consistency of raw materials. It should be stressed that there is a requirement of processors also, that they should improve the control of their processing operations, because without that improvement the effect of improved quality and consistency of oils, fillers, rubbers, etc., would be lost.

There are three contributing factors to inconsistency of the products of mixing; they are variation in raw material quality, poor control of the process, and the use of empirical test methods. These will be considered separately, but to ensure an adequate, comprehensive quality system requires that it be judged against some authoritative standard. Three examples of such standards are: British Standard PS5750 on Quality Systems, the NATO Quality Control System for Industry AQAP-4, and the US Military Specification MIL-Q-9858A. These are essentially similar to one another, and cover all the components necessary for evaluating the capability of a quality management system. Probably the best check of compliance with such standards is the production of a quality manual for internal use, which ensures that all employees are aware of the procedures to be followed to ensure that the company's products are of the required quality.[85]

5.2. Raw Material Quality

The major raw materials of the rubber industry are, in addition to the rubber itself, carbon black, other fillers, oils, and various chemicals. For each of the categories the processors have to establish the minimum requirements for suppliers. These minimum requirements are usually stated in such terms as 'enough to ensure that suppliers have a quality system to assure that all material supplied conforms to all quality requirements.'[86] This, in effect, means that the suppliers must have in place a quality system covering raw materials, processes, and also production.[87] As Haverhals[88] says, 'When confidence is gained that guaranteed quality is being provided, incoming raw material testing can be relaxed and inventories minimized.'

In defining the quality required for any particular application it is essential to determine the relative importance of quality, performance and price. This can only be done by maintaining close communication between the supplier and the customer. This enables development of specifications and standards that are both right for the customer, and feasible for the producer.

Specifications usually include minimum and/or maximum property limits, both for the raw elastomer and for specified compounds. There are also certain implicit, but unstated, properties that are understood by both producer and customer to be part of the agreement. An obvious example is that the product should be free of contamination by foreign materials.

The purpose of specifications is frequently misunderstood. They do not, in themselves, guarantee that the rubber will mix, mould, cure or perform satisfactorily in the customer's process or product. In many cases the test recipe bears little or no resemblance to the compound that the customer produces from the rubber. Equally, the tests that are performed on the raw rubber or compound, and which feature in the specification, often bear little resemblance to the actual processing or service requirements. The object or purpose of specifications, as applied to elastomers, is to guarantee to the customer that he receives that same product shipment by shipment. Thus the specification is, in effect, a benchmark or template. If the rubber meets the agreed specification then it can be expected to perform in processing and application in a consistent manner. The specification is thus a measure of the consistency of the product.

The width of a specification range for a particular property is also important, and it must reflect the customer's needs and the manufacturer's capability to produce and supply.

Thus, in total, the specification should be seen as a statistical limit, about a mid-point, of variation of a property. The mid-point reflects the customer's needs; the limits reflect both the customer's needs and the manufacturer's capability. Its main purpose is to guarantee consistency of the rubber from shipment to shipment.

5.3. Process Control

Control of the mixing process has already been discussed earlier in this review, where it was stated that it is difficult to see how the necessary degree of consistency could be achieved other than by microprocessor control. The use of the statistical tools, referred to above and described in detail by Haverhals,[88] to control a mixing process with batch times of 10 min or less is not easy. However, they can be used to monitor the process and, especially, the product of mixing. Once again quoting Haverhals,[88]

'Recording and analyzing the data will establish whether the process is in (statistical) control and therefore predictable.

If this is indeed the case, process capabilities can be calculated. Process capabilities indicate the variability (precision) and centering (accuracy) of the process output against the target and as a consequence, the percentage production which does not or cannot meet the product specification.'

The quality demands on rubber processors today originate mostly from the automotive industry. Most automobile manufacturers insist that their suppliers provide proof that their processes are under statistical control. This proof may be by actual inspection, audit of the supplier's factory or by insistence that the supplier show evidence that statistical process control (SPC) techniques are effectively employed in his plant. This process is carried one step further back by the moulders to their suppliers, the manufacturers of carbon black, rubber, oils, plasticizers, additives, etc.

The statistics involved in SPC are elementary, little more than an understanding of the basic mathematics of the Gaussian distribution. The actual techniques used, the application of the statistics, are not new. '$\bar{X}$' and 'R' charts have existed for at least 50 years. The belated realization that such simple techniques work is new. What is also new is the added realization that quality is not the concern of the Quality Control Department alone; that quality is, in fact, the concern of everyone.

There are two other important points. The first is that poor quality costs money. Although the adoption of SPC and other tools may initially require investment, the overall result of improving quality is beneficial financially. The second point is that top management has to be involved and committed. In general, the workers can correct (that is, be responsible for) not more than 15% of the variation in a product. If the raw materials, tools, or procedures are inadequate, they can do nothing about them. The corollary is that the other 85% of the variation is the responsibility of management. If they do not act, this 85% remains uncorrected.

5.4. Test Methods

One of the difficulties in the rubber industry is that many of the established and accepted test methods are empirical. An example of this is the Mooney test. As a result there is no such thing as an 'absolute' value of the Mooney of a sample of rubber, in the way that there is an absolute value of its density. Therefore, careful calibration of the instruments, the use of reference, or standard, materials and monitoring the performance of the equipment with control charts, are essential.

6. SUMMARY

The intent of this review was to describe the developments in rubber mixing since Palmgren's paper,[1] and to show how well the needs which he pointed out have been answered. In addition, it has drawn attention to developments in microprocessor control, and the need for increased consistency, or quality, of the product of mixing.

Our understanding of the mixing process has improved, and will continue to improve, as will the ability to control the mixing process. In addition, it is an easy prediction, because it is already well underway, that improvements in quality will be the single, most important theme of the rubber industry for the next decade.

REFERENCES

1. PALMGREN, H., (a) *Europ. Rubber J.*, **156**, 1974, 30; also reprinted in (b) *Rubber Chem. Technol.*, **48**, 1975, 462.
2. FUNT, J. M., *Mixing of Rubbers*, RAPRA Publications, Shawbury, UK, 1977.
3. MORRELL, S. H., *Prog. Rubber Technol.*, **41**, 1978, 97.
4. JOHNSON, P. S., in *Basic Compounding and Processing of Rubber*, ed. H. Long, Rubber Division, ACS, Washington, 1985.
5. HINDMARCH, R. S. and GALE, G. M., (a) *Europ. Rubber J.*, **164**, 1982, 29; also published in (b) *Elastomerics*, **114**(8) 1982, 20.
6. TOKITA, N. and PLISKIN, I., *Rubber Chem. Technol.*, **46**, 1973, 1166.
7. NAKAJIMA, N., *Rubber Chem. Technol.*, **53**, 1980, 1088.
8. NAKAKIMA, N., *Polym. Eng. Sci.*, **19**, 1979, 215.
9. NAKAJIMA, N., *Rubber Chem. Technol.*, **54**, 1981, 266.
10. NAKAJIMA, N., *Rubber Chem. Technol.*, **55**, 1982, 931.
11. COTTEN, G. R., *Rubber Chem. Technol.*, **54**, 1981, 61.
12. FREAKLEY, P. K. and WAN IDRIS, W. Y., *Rubber Chem. Technol.*, **50**, 1977, 163.
13. NAKAJIMA, N., and HARRELL, E. R., *Rubber Chem. Technol.*, **56**, 1983, 197.
14. GESSLER, A. M., HESS, W. M. and MEDALIA, A. I., *Plastics and Rubber Processing*, **3**, 1978, 109.
15. EBELL, P. C. and HEMSLEY, D. A., *Rubber Chem. Technol.*, **54**, 1981, 698.
16. HESS, W. M., SWOR, R. A. and MICEK, J. E., *Rubber Chem. Technol.*, **57**, 1984, 959.
17. BOONSTRA, B. B. and MEDALIA, A. I., *Rubber Chem. Technol.*, **36**, 1963, 115.
18. DIZON, E. S., MICEK, E. J. and SCOTT, C. E., *J. Elastomers Plast.*, **8**, 1976, 414.
19. DIZON, E. S., *Rubber Chem. Technol.*, **49**, 1976, 12.
20. McKELVEY, J. M., *Polymer Processing*, Wiley, New York, 1962, pp. 326–32.
21. TADMOR, Z., *Ind. Eng. Chem. Fundam.*, **15**, 1976, 346.
22. MANAS-ZLOCZOWER, I., NIR, A. and TADMOR, Z., *Rubber Chem. Technol.*, **55**, 1982, 1250.

23. COTTEN, G. R., (a) *Rubber Chem. Technol.*, **57**, 1984, 118; (b) *Rubber Chem. Technol.*, **58**, 1985, 774; (c) *Kautsch. u. Gummi Kunst.*, **38**, 1985, 705.
24. MANAS-ZLOCZOWER, I., NIR, A. and TADMOR, Z., *Rubber Chem. Technol.*, **57**, 1984, 583.
25. NAKAJIMA, N. and HARRELL, E. R., *Rubber Chem. Technol.*, **57**, 1984, 153.
26. SHIGA, S. and FURUTA, M., *Rubber Chem. Technol.*, **58**, 1985, 1.
27. TURETSKY, S. B., VAN BUSKIRK, P. R. and GUNBERG, P. F., *Rubber Chem. Technol.*, **49**, 1976, 1.
28. FREAKLEY, P. K. and PATEL, S. R., *Rubber Chem. Technol.*, **58**, 1985, 751.
29. ELLWOOD, H., *Europ. Rubber J.*, **159**, 1977, 17.
30. WHITAKER, P., *Kautsch. u. Gummi Kunst.*, **34**, 1981, 295.
31. WIEDMANN, W. M., SCHMID, H.-M. and KOCH, H., *Kautsch. u. Gummi Kunst.*, **33**(11), 1980, 926.
32. WIEDMANN, W. M. and SCHMID, H.-M., *Kautsch u. Gummi Kunst.*, **34**(6), 1981, 479.
33. WIEDMANN, W. M. and SCHMID, H.-M., *Europ. Rubber J.*, **164**, 1982, 33.
34. WIEDMANN, W. M. and SCHMID, H.-M., *Rubber Chem. Technol.*, **55**, 1982, 363.
35. SCHMID, H.-M., *Rubber World*, February 1984, 33.
36. O'CONNOR, G. E. and PUTNAM, J. B., *Rubber Chem. Technol.*, **51**, 1978, 799.
37. DOLEZAL, T. P. and JOHNSON, P. S., *Rubber Chem. Technol.*, **53**, 1980, 252.
38. JOHNSON, P. S., *Kautsch u. Gummi Kunst.*, **33**, 1980, 725.
39. WOLFF, S., *Rubber World*, June 1984, 28.
40. BASIR, K. B. and FREAKLEY, P. K., *Kautsch. u. Gummi Kunst.*, **35**, 1982, 205.
41. MEISSNER, K. and REHER, E.-O., *Plaste u. Kautsch*, **27**, 1980, 514.
42. STUDEBAKER, M. L. and BEATTY, J. R., *Rubber Age*, **103**(5), 1976, 21.
43. TOPCIK, B., *Rubber Age*, **105**(7), 1973, 25; *ibid.*, **105**(8), 1973, 35.
44. TANAKA, V., *Nippon Gomu Kyokaishi*, **7**, 1981, 437; translated in *Int. Polym. Sci. Technol.*, **9**(1) 1982, T186.
45. TADMOR, Z. and MANAS-ZLOCZOWER, I., *Rubber Chem. Technol.*, **57**, 1984, 48.
46. HOWGATE, P. G. and COTTEY, D. P., *Prog. Rubber Technol.*, **43**, 1980, 99.
47. RAPETSKI, W., *Elastomerics*, September 1981, 21.
48. ACQUARULO, L. A. and NOTTE, A. J., *Elastomerics*, November 1983, 17.
49. MULLER, D. and MAIRE, U., *Rubber World*, February 1984, 25.
50. SYMPOSIUM, *Applications of Microprocessors in the Rubber Industry*, ACS Rubber Division, 123rd Meeting, Toronto, May 1983.
51. DIZON, E. S. and PAPZIAN, L. A., *Rubber Chem. Technol.*, **51**, 1978, 799.
52. THIBODEAU, W. E., JOHNSON, P. S. and GARK, K. R., Paper presented at the Symposium ref. 50; published in *Rubber World*, **188**(1), 1983, 21.

53. NORMAN, R. H. and JOHNSON, P. S., *Rubber Chem. Technol.*, **54**, 1981, 493.
54. MIERAS, H. J., International Rubber Conference, Brighton, UK, May 1972.
55. NAKAJIMA, N. and COLLINS, E. A., *Trans. Soc. Rheol.*, **20**(1), 1976, 1.
56. DUNN, J. R., WOOD, A. G. and BYRNE, P. S., Interscandinavian Rubber Meeting, Copenhagen, May 1976.
57. SMITH, B. R., *Rubber Chem. Technol.*, **49**, 1976, 1316.
58. NAKAJIMA, N. and COLLINS, E. A., *Rubber Chem. Technol.*, **47**, 1974, 333.
59. NAKAJIMA, N., and HARREL, E. R., *Rubber Chem. Technol.*, **52**, 1979, 962.
60. NAKAJIMA, N., and HARREL, E. R., *Rubber Chem. Technol.*, **55**, 1982, 1426.
61. TURNER, D. M. and MOORE, M. D., *Plastics and Rubber Processing*, **5**, 1980, 81.
62. TURNER, D. M. and BICKLEY, A. C., *Plastics and Rubber Processing and Applications*, **1**(4), 1981, 357.
63. PICA, D., BARKER, R. I., RICE, P. and MA, C. C., *Rubber World*, **180**, July 1979, 95.
64. EINHORM, S. C. and TURETSKY, S. B., *J. Appl. Polym. Sci.*, **8**, 1963, 1257.
65. LEBLANC, J. L., *Plastics and Rubber Processing and Applications*, **1**, 1981, 187.
66. BITTEL, P. A., *Elastomerics*, April 1980, 44.
67. SEZNA, J., *Rubber World*, November 1983, 27.
68. SEZNA, J., Paper 4, Symposium on *Elastomer Processing*, ACS Rubber Division, 128th Meeting, Cleveland, USA, October 1985.
69. KARG, R. F., *Rubber World*, **187**, 1983, 28.
70. BERRY, J. P., SAMBROOK, R. W. and BEESLEY, J., *Plastics and Rubber Processing*, **2**, 1977, 97.
71. MILLS, M. and GILLESPIE, R. M., 19th Annual Meeting of the IISRP, Hong Kong, April 1978.
72. LEBLANC, J., *Europ. Rubber J.*, **162**(1), 1980, 20.
73. AMSDEN, C. S., *Polymer Testing*, **5**(1), 1985, 45.
74. HESS, W. H. and WIEDENHAEFER, *Rubber World*, September 1982, 15.
75. MEDALIA, A. I., Educational Symposium on *Efficient Rubber Mixing Technology*, ACS Rubber Division, 118th Meeting, Cleveland, USA, October 1981.
76. PERSSON, P. S., *Europ. Rubber J.*, November 1978, 28.
77. (a) VEGVARI, P. C., HESS, W. M. and CHIRICO, V. E., *Rubber Chem. Technol.*, **51**, 1978, 817: (b) *idem*, US Patent 4229042, 2 September, 1980; (c) HESS, W. H., CHIRICO, V. E. and VEGUARI, P. C., *Elastomerics*, **112**(1), 1980, 24.
78. DEN OTTER, J. L. and GERRITSE, G. A. *Plastics and Rubber Processing and Applications*, **4**, 1984, 63.
79. Product Quality Symposium, IISRP Annual General Meeting, Den Hague, 7–11 May, 1984.

80. *Symposium Quality Opportunities for the 80s,* 126th Meeting, ACS Rubber Division, Denver, Colorado, USA, 25 October, 1984.
81. JURAN, J. M., *Management Review,* June 1981, 9 and July 1981, 57.
82. LONG, J., *Basic Statistical Process Control,* paper published by the Ontario Center for Automotive Parts Technology, Canada.
83. SIEGEL, J. C., SAE Technical Paper 820520, International Congress and Exposition, Detroit, Michigan, 22–26 February, 1982.
84. ANON., *Rubber World,* January 1985, 15.
85. WAIN, B. J., Rapra Members Report No. 85, 1983.
86. A paraphrase of *Firestone's General Policy for Supplier Quality Requirements,* Firestone World Tire Group, Akron, Ohio, USA, 1984.
87. FRANCIS, A. E., Paper at CIC Meeting, Waterloo, Ontario, 16 October, 1984.
88. HAVERHALS, L., *Quality Improvement and Quality Assurance,* Polysar, obtainable from Polysar Technical Centre NV, PO Box 354, B-2000, Antwerp, Belgium, 1985.

Chapter 8

HEALTH AND SAFETY

B. G. WILLOUGHBY

Rapra Technology Ltd, Shawbury, Shrewsbury, UK

NOTATION

a	radius
A	cross-sectional area
C	root mean square (RMS) speed
C_d	drag coefficient
d	diameter
D	diameter; coefficient of diffusion
$\mathbf{F}$	frictional force
g	acceleration due to gravity
L	molar latent heat of evaporation
m	mass
M	molecular weight
$\mathbf{N}$	normal force
P	pressure
Q	volume rate of air flow
R	gas constant: $R = 8\cdot31 \times 10^7$ erg/K mol or $R = 8\cdot21 \times 10^{-2}$ litre atm/K mol
R_e	Reynolds number: $R_e = dv\sigma/\eta$
T	Kelvin temperature
$\left.\begin{array}{c}u \\ U \\ v\end{array}\right\}$	velocity
V	velocity, volume

v_c	terminal velocity
η	coefficient of viscosity: for air at 20°C and 760 mm Hg, $\eta = 18 \cdot 2 \times 10^{-6} \, \text{Ns/m}^2$
μ	coefficient of friction
ρ	particle density
σ	fluid density

ABBREVIATIONS

ACGIH	American Conference of Governmental Industrial Hygienists
BP	boiling point
BRMA	British Rubber Manufacturers' Association
DS	dust suppressed
GC	gas chromatography
GC/MS	gas chromatography with on-line mass spectrometry
GTM	gas transfer mould
HSE	Health and Safety Executive
LEV	local exhaust ventilation
LP	liquid petroleum
OEL	occupational exposure limit
ppm	parts per million
RMS	root mean square
SG	specific gravity
TWA	time-weighted average
VHI	vapour hazard index

CHEMICAL ABBREVIATIONS

ADC	azodicarbonamide
AZDN	azobisisobutyronitrile
BR	polybutadiene rubber
BTPPC	benzyltriphenylphosphonium chloride
CBS	N-cyclohexylbenzothiazyl-2-sulphenamide
DBP	dibutyl phthalate
Dicup	dicumyl peroxide
DPG	diphenylguanidine
DTBP	di-tert-butyl peroxide

DNPT	dinitrosopentamethylenetetramine
EPDM	ethylene–propylene terpolymer
FKM	fluorocarbon rubber
IPPD	N-isopropyl-N'-phenyl-p-phenylenediamine
MBOCA	4,4'-methylenebis(2-chloroaniline)
MBT	2-mercaptobenzothiazole
MBTS	2,2'-dibenzthiazyl disulphide
Me	methyl group
MIBK	methyl isobutyl ketone
NBR	nitrile rubber
NDPA	N-nitrosodiphenylamine
NOBS	N-oxydiethylene-2-benzothiazyl sulphenamide
NOX	oxides of nitrogen
NR	natural rubber
Ph	phenyl group
SBR	styrene–butadiene rubber
TBTD	tetrabutylthiuram disulphide
TMTD	tetramethylthiuram disulphide
6PPD	N-1,3-dimethylbutyl-N'-phenyl-p-phenylenediamine
77PD	N,N'-bis(1,4-dimethylpentyl)-p-phenylenediamine

1. INTRODUCTION

The avoidance of illness or injury is inevitably a subject of universal concern, yet even the recognition of potential hazards may involve confronting issues of great complexity. For example, a slipping accident might be correctly addressed as a lifting problem: powered transportation could provide one solution but introduce new hazards if exhaust emissions directly or indirectly cause unacceptable pollution. Problems of this nature may not be unique to the rubber industry, but the interdependence of numerous variables reflects much of the general character of this industry. Achieving fast cycle times without scorch, and balancing conflicting trends in product performance, are routine activities of a manufacturing operation which is overtly engineering and covertly chemical. Such experience must provide a useful foundation for gaining an appreciation of the factors behind workplace hazards and this chapter aims to provide some basic elements of the understanding now emerging from current research. Without doubt our understanding in these matters is far from perfect

and some attempted rationalizations might be more safely regarded as
preliminary. Indeed, much may remain unresolved without a wider
contribution of experience to these issues.

Before proceeding further, it will be worthwhile to set limits on the
task in hand. The maze of relevant standards or codes of practice in
operation worldwide is a subject as complex as it is important. A
recent and able summary by Lawson[1] lifts the veil on these and there
need not be further duplication here. However, the new status of
occupational exposure limits[2] in the UK deserves to be noted, as does
the extension of the BRMA recommended limit on cyclohexane-
soluble fume to include the measurements by personal sampling[3] (see
Section 4) and the planned adoption of this recommendation as an
HSE control limit. Atmospheric sampling itself must be recognized as
a specialist subject, inappropriate for in-depth coverage here.
Comment on specific methods appropriate to the rubber industry is
provided in several sources, including the BRMA Code of Practice[3]
and the work by Nutt,[4] whereas a more general introduction to the
techniques of atmospheric vapour sampling has been compiled by
Thain.[5]

Medical research has been extensive and covers more than three
decades from the recognition of occupational bladder cancer in UK
rubber workers. Other cancers have since been the focus of investigation, as have the patterns for various work-related health problems.
Attention is increasingly being directed to the origins of these
problems and particularly to the physical and chemical environment of
the workplace. This attention provides the principal theme of this
chapter, but for an appreciation of the breadth and detail of relevant
medical research the reader is directed to the reviews listed in the
BRMA Code of Practice[3] and to the eminently readable monograph
by Nutt.[4]

2. ACCIDENTS

Accidents may occur in any human activity, be it in the home, at work
or in a leisure pursuit. Accidents at work have long been the subject of
study and statistics have been regularly compiled, both locally and
nationally. Such figures leave little doubt over the scale of the problem
concerning accidental injury; for example in 1980 more than a quarter
of a million working hours were lost as a result of accidents within the

UK rubber industry. However, the identification and rationalization of trends is more difficult, largely because of the subjective element in the recognition and recording of original data or in the treatment of any injuries sustained, and also because of changes within the industry itself. Within the latter context is the recent contraction of the UK rubber industry, from an activity of 98·2 million man-hours in 1950 to one of 56·5 million man-hours in 1984.[6,7]

2.1. Accident Frequency

One convention which compensates for changes in the scale of the industrial operation is to quote the 'accident frequency', this being the number of accidents per 100 000 man-hours worked. This may be applied either to all accidents or to so-called 'lost-time' accidents, with the amount of time lost appropriately specified. For example, Table 1 gives data for all accidents, those accidents which cause the loss of one shift or more, and those accidents which caused lost time in excess of three days.

The compilation of accident frequencies enables major trends (Table 1) such as their reducing values of recent years, to be readily recognized. Another feature is the slight reduction in the proportion of accidents which are lost-time: 5·7% in 1980 compared with 5·9% in 1977. This downward trend is more apparent when compared with the equivalent figure for a sample survey conducted during 1950–1952, in this case 10·8%.[6] The same trend may be recognized in other statistics, for example as hours lost per lost-time accident (144, 1980; 153, 1977).[7] But progress such as this can only have occurred, and can only be maintained, by firstly recognizing some of the factors which cause

TABLE 1

ACCIDENT FREQUENCY WITHIN THE UK RUBBER INDUSTRY

	1977	1980	1983
Hours worked (millions)	90·65	79·03	60·22
All accident frequency (number of accidents per 10^5 man hours)	44·95	35·50	
Lost-time accident frequency (number of accidents per 10^5 man hours)	2·66[a]	2·01[a]	1·23[b]

[a] Lost time: one shift or more.
[b] Lost time: three days or more.

accidents. This in itself is no mean task and must be considered as an on-going study.

One feature of current practice in the compilation of accident records is the assignment of a principal cause from a standard range of categories. Table 2 gives such data from 1983 and 1984 statistics for the UK rubber industry and also for a sample study in the same industry during 1950–1952. In all cases the figures are for lost-time accidents (lost time: three days or more) and cover total numbers of 764 for 1984, 744 for 1983 and 1565 for 1950–1952. A search for improving trends in these data reveals a significant reduction in the proportions of serious accidents arising from machinery or falling material. With regard to the former, credit must go to improvements in guarding, that to the guarding of rubber processing machinery being an obvious consideration.[8] However, a fuller appreciation of the gains achieved should take a wider perspective: a diverse range of power tools may be encountered in the industry, and the 1950–1952 survey indicated that the largest number of machinery accidents occurred with lathes and buffing machines.[6] That same survey found that more than

TABLE 2

BREAKDOWN OF LOST-TIME ACCIDENTS OVER THREE DIFFERENT PERIODS WITHIN THE UK RUBBER INDUSTRY

Attributed cause of accident/injury	Percentage of lost-time accidents		
	1950–1952	1983	1984
Machinery	15·5	11·0	9·9
Hand-tool	8·9	6·6	4·1
Falling material	19·9	10·5	11·4
Persons falling	18·8	18·7	17·9
Transport	ND[a]	3·9	2·9
Collision	4·9	8·5	7·6
Minor causes	7·6	7·4	7·7
Heat	3·3	0·5	1·3
Corrosive agent	ND	0·5	—
Poisonous substances	ND	0·1	—
Electrical equipment	ND	0·4	—
Explosions	ND	0·1	0·1
Muscular stress	ND	27·0	31·0
Other	21·2	5·1	6·0

[a] ND, no data.

half the injuries caused by falling materials were to the ankle or foot, and clearly worthwhile benefits will have been gained from the routine use of safety footware.

One facet of these results highlights the difficulties in investigative studies of accident statistics, namely that one major factor in producing injury was not even recognized as such in the 1950–1952 study. These are the injuries arising from muscular stress, the most frequent cause of lost-time accidents in the 1983 and 1984 data. These accidents, and those arising from persons falling, account for nearly 50% of the 1983–1984 lost-time accidents summarized in Table 2 and merit more detailed consideration.

2.2. Muscular Stress

Without doubt rubber processing and fabrication is energy-intensive, in both mechanical and human terms. Indeed, the scale and power of the machinery involved may themselves make demands on the workforce in terms of the manual effort to supply or load them with material. But the material itself may make special demands by virtue of such characteristics as elasticity and tack. When rubber being tugged suddenly yields to that pull, then muscular injury should not be unexpected.[9] Veys lists the commonest complaints arising from these labours as muscular strains and sprains, tenosynovitis (inflammation around the tendons) and especially back injuries.

Whereas it may be self-evident that working in the rubber industry is physically demanding, research studies of this field are rarely found. Three studies have been particularly concerned with the convenience of layout within the working zone and with controlling excessive bodily movement.[10–12] Two out of the three were concerned with tyre building and drew special attention to the manual effort involved.[11,12] The use of liners to prevent the blanks sticking together was a significant recommendation.[11] With hindsight such procedures may seem obvious, as might be the introduction of mechanical transport (e.g. fork-lift trucks) to minimize manual lifting and carrying. But it is evident from the scale of the problem (Table 2) that a considerable scope for improvement remains. It may be that in this case the first step to worthwhile gains will be a wider appreciation of the nature of the problem. Beyond this the implications of options for improvement should also be considered, particularly if mechanized alternatives can introduce a different range of hazards.

2.3. Falling Bodies—Persons Slipping

Injuries sustained in falling occur on a scale to merit widespread concern: UK statistics for all such accidents—at work or elsewhere—indicate that falls in 1979 produced more than half a million serious injuries and nearly 6000 fatalities.[13] Within the manufacturing industries that year there were 38 321 falls which caused an absence from work of more than three days and 30 fatalities. It is falls from a height (from ladders, stairs, scaffolding, etc., or from one level to another) which most commonly cause death,[13] and these are atypical of the majority of falling accidents in the rubber industry.[6,8] Indeed, within this industry fatal accidents from any cause are thankfully rare.[8]

Statistics for the 1950–1952 period revealed that 74% of the falls studied occurred at floor level;[6] for the period 1975–1977 such falls accounted for 58% of the total reported.[8] When such accidents occur as a result of obstructions or uneven floor surfaces, preventative action is obvious, but the prevention of slipping falls—once shown[6] to be the majority type—is less straightforward. Not surprisingly pedestrian friction is now emerging as a key area for research study.[14–16]

Fundamental to these studies are the classical laws of friction,[16] and the basic relationship (eqn (1)) between frictional force (**F**) and normal force (**N**).

$$F = \mu N \tag{1}$$

For walking, the frictional force at the ground must at least match the horizontal thrust exerted by the foot to achieve the required acceleration.[14] For running, the acceleration is higher and correspondingly higher levels of friction are required. The constant of proportionality in eqn (1) is the coefficient of friction (μ) and is dependent on both the materials in contact, e.g. the shoe sole and the floor. Clearly the magnitude of μ determines what friction can be generated and what motion can be sustained: for example a value of 0·5 might be adequate for walking, but values in excess of 1·0 may be necessary for running.

Most recent research in this area has focused on the factors which can cause variations in either μ or **N** in eqn (1). Grieve[17] for example considers the effect of manual exertion on normal force. The exertion in work can change the 'effective weight' of a person and such changes can be quantified by means of a simple vector analysis (Fig. 1). The direction of exertion emerges as the key factor, the greatest hazard occurring when exertion has a downward component. Similarly the

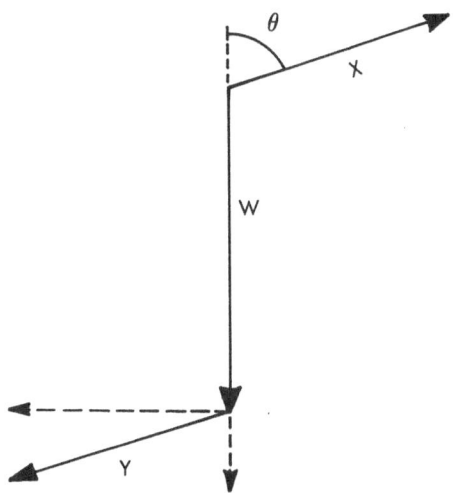

FIG. 1. Vector analysis showing the effect of manual exertion on normal force. In this case the body weight is represented by the vector **W** and the effort, applied at an angle θ to the vertical, is represented by a vector **X** and opposed by a reaction force at the ground shown as the vector **Y**. Thus the total downward force is increased to **W** + **Y** cos θ and the frictional force required for stability is **Y** sin θ. From eqn (1), $\mu =$ **Y** sin $\theta/($**W** + **Y** cos $\theta)$.

magnitude of that exertion is important: the greater this is, then the greater the likelihood that an unstable situation may develop.

James[16,18] focuses attention on the factors which influence the magnitude of μ, but firstly sets out to rationalize which of two coefficients of friction—static or kinetic—is critical in this context. The question of stability he views not as an exercise in statics, but as a situation governed by the consequences of effecting a slight change in the *status quo*. Stability emerges as a dynamic problem, one of arresting a movement rather than one of guaranteeing the absence of motion, a concept which now gains experimental support from high-speed photography.[19] By this argument the condition for stability can no longer be described in terms of a single value of μ, but more properly in terms of the requirement that μ must increase with increasing velocity.

This strict adherence to practicalities may negate the treatments of simple physics, but it does offer the rubber scientist opportunities for a closer appreciation of current research. For example rubber materials are ubiquitous in shoe soling, and rubber friction has been extensively studied, with contributions from adhesive and deformative mechanisms now recognized.[20,21] When a dynamic parameter is under consideration then a marked compound and temperature dependence will be expected. James illustrates this point by reference to a polyurethane[16] which is satisfactory, and to a nitrile rubber[18] which may not be so (Fig. 2). Where adhesion plays a part in friction then surface contamination must be considered, and again experiment[16] reveals that marked changes may result (Fig. 3).

The understanding of this topic which is now emerging recognizes the need for more reference data, but it is clear that the right rubber product can play an important role in workplace safety. Further studies should also take account of the true nature of the surfaces involved, which with regard to the floor itself can include coatings of dust, detergent residues, polish, contamination through spillages or combinations of these. If satisfactory conditions can be created, then they must be maintained—a requirement which places challenges at all levels and with respect to both investigation and organization.

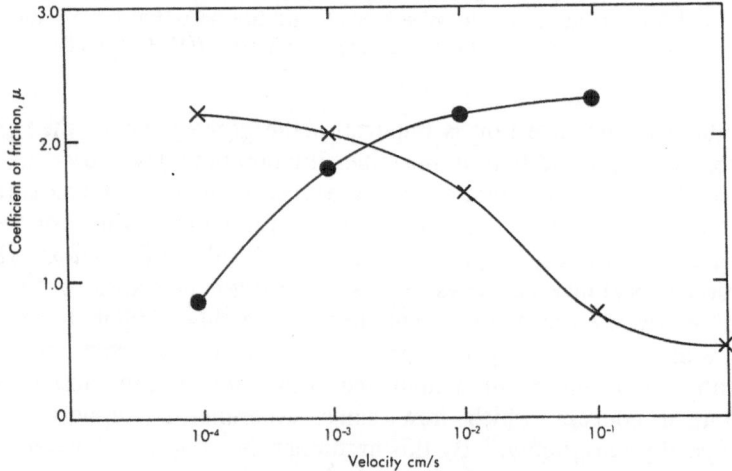

FIG. 2. Coefficients of kinetic friction for NBR sliding against glass at two different temperatures. At 20°C (●) the situation is stable with μ already at a high value at low velocities and rising sharply as velocity increases. At −5°C (×), μ is higher than the 20°C value at low velocities but the situation is no longer stable since μ now falls markedly as velocity increases. (Reproduced by courtesy of Taylor and Francis Ltd, UK.)

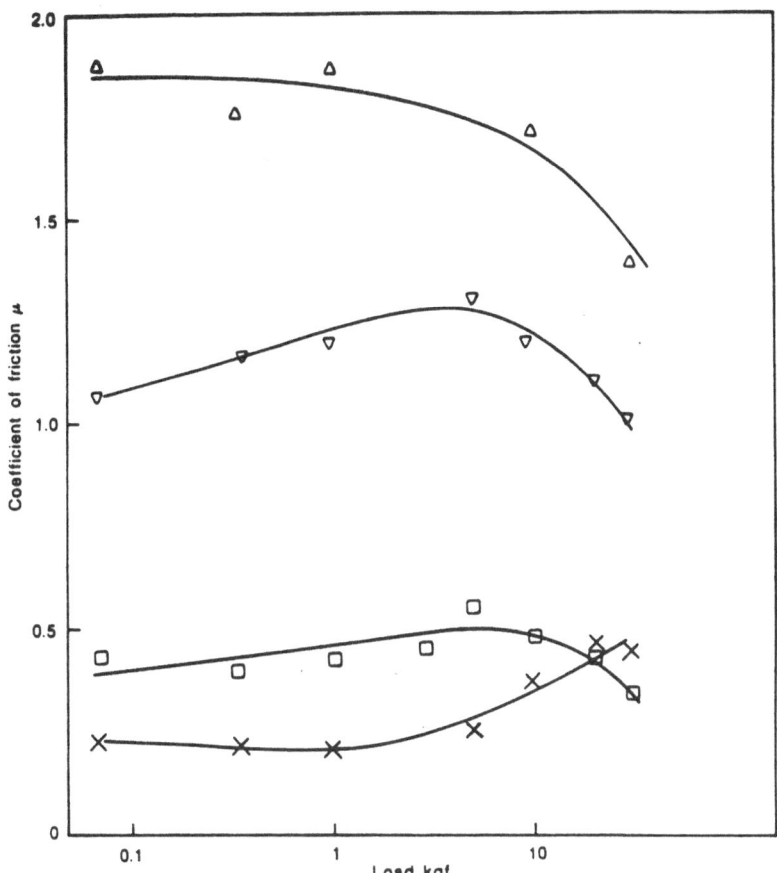

FIG. 3. Coefficient of kinetic friction for plasticized PVC on steel under different conditions of contamination.[16] In all cases the PVC was plasticized with 60% DBP and μ measured at 0·3 cm/s under dry ($\triangle$), dusty ($\square$) and wet ($\triangledown$) conditions, and also in the presence of both dust and water ($\times$). Whereas water significantly reduces friction, the effect of the dust or slurry is markedly greater. (Reproduced by courtesy of Rapra Technology Ltd, UK.)

3. DUST HAZARDS

The rubber industry has traditionally handled a diverse range of powders, many of which pose some threat to health. For example the BRMA 1985 Code of Practice[3] lists more than 50 organic and inorganic solids as the 'Category B' chemicals which merit extra care in handling. Amongst these are examples from all types of rubber chemicals including many of the common accelerators and all of the blowing agents catalogued. Additionally a 'special category' of hazard is reserved for chemicals, such as ethylene thiourea, tellurium diethyldithiocarbamate and MBOCA. Formerly the preferred form[22] for solid additives was as fine powders which could be readily dispersed in rubber, and retrospective mortality studies have produced evidence for one occupational disease which might result from exposure to dust in the rubber industry.[23] Not surprisingly therefore, the dust content of workplace air has been the subject of concern and study.[8,24,25] Inevitably considerable attention has been focused on reducing airborne dust concentrations in the workplace.

3.1. Falling Bodies—Particles in Air
To gain an insight into how such reductions can be achieved it is important to consider how dust clouds are generated, and how they move or can be moved. Much of this pollution can be clearly seen, and the picture of particles apparently floating in the air may be one impression gained. However, floating in the classical sense is certainly not possible since buoyancy offered by the displacement of air by the particle would be insufficient to support the weight of that particle (Fig. 4). This absence of buoyancy means that the motion of the particle is likely to have a measure of independence of that of the air around it, a characteristic which must have profound implications for dust control.

3.1.1. Particle Dynamics
The manner by which the air or its flow can influence the motion of a particle is described by particle dynamics, a subject of widespread study and considerable complexity.[26-28] The principal influence of the surrounding fluid (air) on particulate motion is by means of the frictional forces which will be generated to oppose the relative motion of the particle with respect to that fluid. This frictional force (F) depends on the density of the fluid (σ) and the velocity (v) of the

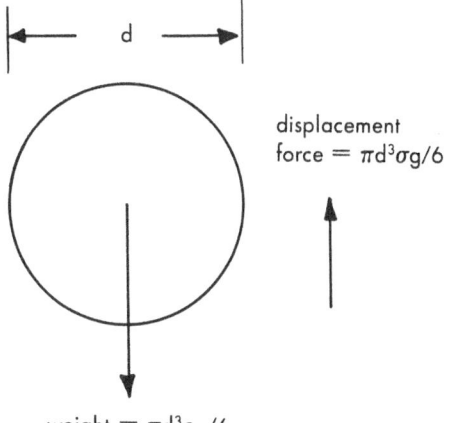

FIG. 4. Forces acting on a stationary sphere of density ρ in a fluid of density σ. When the fluid is air the weight is likely to exceed the displacement force by a factor of 10^3 or more.

particle and its cross-sectional area (A), in a plane normal to the direction of travel, as given by eqn (2).

$$F = C_d A \sigma v^2 / 2 \qquad (2)$$

The factor C_d is the drag coefficient, which is not a constant, but which actually decreases in value (perhaps by orders of magnitude[28]) as particle size and velocity increase. There is a relationship, albeit complex,[26,28] between C_d and the dimensionless Reynolds number (defined for this system as eqn (3)[29]), and for spherical particles of diameter d, algebraic relationships exist.

$$R_e = \sigma v d / \mu \qquad (3)$$

Thus for streamline flow with $R_e < 0.2$, eqn (4) applies: for larger particles and increasing turbulence for which $0.2 < R_e < 500$, the empirical eqn (5) is approximately true.[29]

$$C_d = 24 / R_e \qquad (4)$$

$$C_d = 24 \, (1 + 0.15 \, R_e^{0.687}) / R_e \qquad (5)$$

Other equations have also been postulated to cover non-streamline flow and systems of even higher R_e, but eqn (4) is adequate for free-falling airborne spheres in the 2–60 μm range.[26]

3.1.2. Particle Size

For flow in the viscous region, eqns (2), (3) and (4) can be combined to give the familiar Stokes equation (eqn (6)).

$$F = 3\pi\eta\sigma d \tag{6}$$

If the flow is a vertical descent under the action of gravity, this frictional force can act in concert with the hydrostatic displacement force (Fig. 4) so that a terminal velocity (v_c) is achieved. The resulting expression is eqn (7), for a sphere of density ρ, and this forms the basis for the graphical presentation in Fig. 5.

$$v_c = (\rho - \sigma)gd^2/18\eta \tag{7}$$

Particle size emerges as a critical parameter in eqn (7) ($v_c \propto d^2$); thus if the settling distance is 1 m, a 50 μm-diameter sphere of SG 1·0 will cover that distance at its terminal velocity in 13 s whereas a 10 μm diameter sphere of the same SG will require 5·5 min. Dorman[26] has shown the change in v_c with d outside the viscous range (R_e up to 10^4): for a 1 mm-diameter sphere of SG 1·0 the terminal velocity at atmospheric pressure and 20°C is around 40 m/s. Under comparable conditions a 0·1 μm-diameter sphere has a terminal velocity of approximately 3 mm/h.[26] Particle size therefore emerges as a dominant factor in the dynamics of airborne particles, and only the very smallest particles are capable of apparently motionless suspension in still air.

3.1.3. Particle Projection

The effects of air resistance govern any relative motion of a particle with respect to the air surrounding it. For a group of particles projected from a given source, small particles may travel appreciable distances.[26]

The concept of *projecting* a particle merits emphasis, since a denser-than-air particle can only be rendered airborne as a result of a motion imposed upon it. Such motion could be initiated by a movement of the air itself, but commonly in the factory it arises from a mechanical displacement operation (buffing, bag-emptying , walking, etc.). Even when powders are being mixed into rubber, the conditions of high shear and slow wetting by the polymer can still provide opportunities for dust generation. In one extensive monitoring exercise covering up to ten UK rubber factories, the weighing, milling or internal mixer loading areas were consistently amongst the most abundant sources of airborne particulates.[24] In the case of the internal

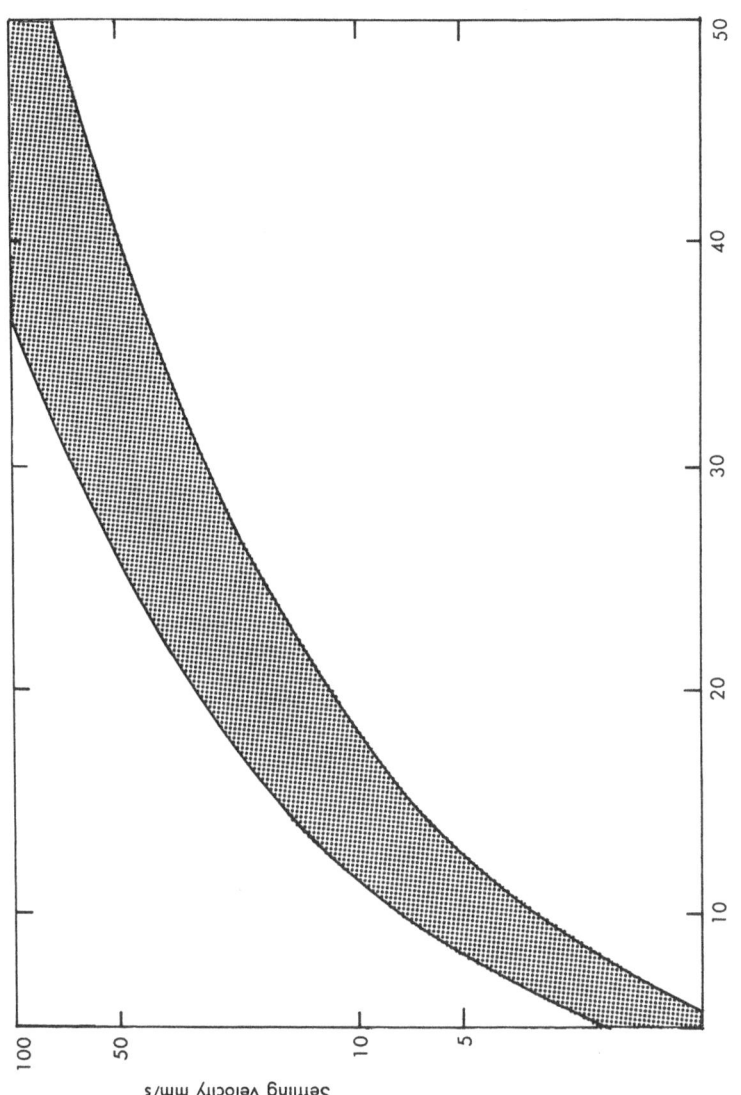

FIG. 5. Range of settling velocities for spheres of specific gravity 1·0–2·5. The band shown in the figure is defined by the v_c versus d curves for SG 1·0 over the diameter range 6–50 μm and for SG 2·5 over the range 5–36 μm.

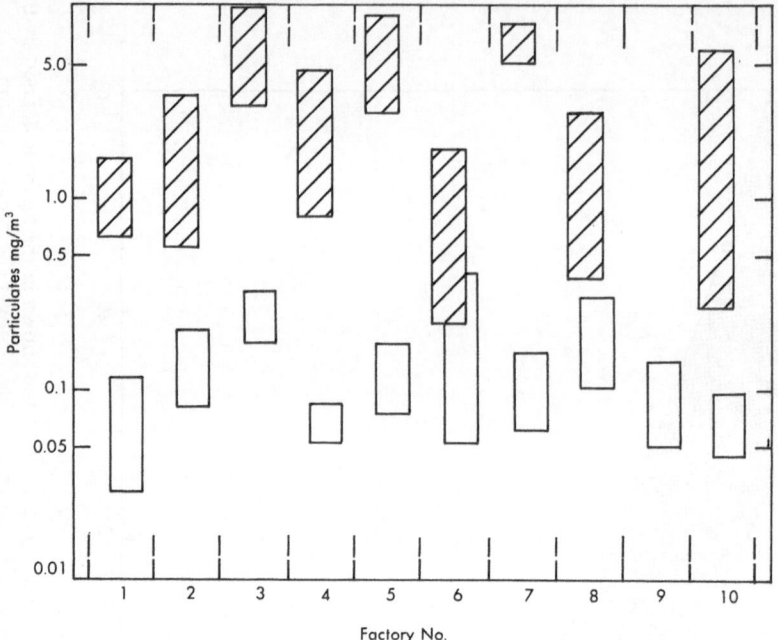

FIG. 6. Atmospheric dust measured in the internal mixer loading areas in ten UK rubber factories. The blocks shown represent the range of values obtained during a period substantially covering the calendar year 1974. The hatched blocks designate the results for the factory site in question and the unhatched blocks are for measurements of the local outside air. (Reproduced by courtesy of the BRMA, UK.)

mixer an additional source of dust is the imposed motion in the air expelled by the ram; the results for one monitoring exercise in this area are summarized in Fig. 6.

However, the industry should not neglect the diverse range of mechanical operations in other process areas which can also generate dust. Many may be familiar engineering practices and share problems in common with other industries. However, even amongst these are operations specific to the rubber industry with unusual dust-generating capabilities; for example, tyre buffing with an 11 in-diameter rasp operating at 2850 rpm can generate particles moving at over 90 miles/h.

3.2. Deflection and Capture of Airborne Particulates

It is clear from considerations of particle dynamics that a cloud of airborne dust can be composed of particles behaving in different ways from one another, moving with different velocities and even in different directions. This and the ability of many particles to travel substantially independently of prevailing air currents pose considerable problems for the implementation of external controls. For example, if particles are travelling within a stream of air which is suddenly deflected there is the likelihood that some of these particles will not be deflected. The system must rely on air resistance to divert the particles from their original path and Dorman has treated the situation for motion in the viscous region by considering components of motion in two directions, i.e. along and transverse to the original direction of travel; e.g. eqn (8) applies to the x-direction.[26]

$$m\mathrm{d}^2x/\mathrm{d}t^2 = 3\pi\eta d(u - \mathrm{d}x/\mathrm{d}t) \qquad (8)$$

where m is the mass of the particle and u the resolute of the air velocity.

3.2.1. Inertia Capture

If the deflection of the air is caused by a rod of diameter D lying transverse to the flow, then the equations of motion can be expressed in terms of the air velocity U remote from the rod and an inertial parameter K defined by eqn (9).

$$K = \rho d^2 U/9\eta D \qquad (9)$$

The inertial parameter shows that the likelihood of impact by a particle on the rod increases with the size and velocity of the particle and with decreasing diameter of the rod (i.e. the severity of the deflection).[30] This might be regarded as the simplest model for the capture of dust on an element of a filter. For air sampling on filters velocities are such that the main mechanism of capture is inertia.[31] By this route the efficiency towards capture of the smallest particles becomes critically dependent on velocity; one standard test achieved 89% penetration of a paper filter by a 0·3 μm aerosol at 0·025 m/s, but only 15% penetration at 1 m/s. For routine factory monitoring sampling speeds as high as 0·7 m/s have been employed,[24] but speeds of less than 0·05 m/s may be encountered if portable battery-operated pumps are used.

It should not be assumed that inertial effects are the only mechanisms contributing to particulate capture on filters, or that the capture efficiency of a filter paper or mat is a simple extrapolation of that of a single fibre.[31] However, the flow rate will always have a bearing on performance, and the avoidance of structural damage to the filter (either in mounting or in use) cannot be over-emphasized.

3.2.2. Inhaled or Ingested Particles

Inertial effects are exploited in other sampling methods, for example in impactors or impingers,[31] but may be expected to play different roles in the segregation of particles being inhaled or ingested within the human body. In the first case the removal of particles from inhaled air will exclude the large particles (e.g. $d > 7 \, \mu m$)[4] from entering the lungs, a sequence which has been extensively described.[4,32,33] In the other case the larger particles thus trapped become available for transfer to the stomach, ultimately by the ingestion of mucus and saliva.[4,34] The existence of an upper limit on the size of particle entering the mouth will depend on many factors, including for example the orientation of the head with respect to ambient air motions. One model study found a 50% interception efficiency for particles as large as 100 μm.[4]

These differences with respect to particulate segregation bear an interesting comparison with differences in apparent trends in cancer mortality amongst rubber workers. For example, some incidence of stomach cancer may be associated with potentially dusty work at the front end of the rubber manufacturing process, but excesses of lung cancer may be connected with later manufacturing stages (e.g. exposure to hot rubber fumes).[4] Thus, on the basis of this evidence, dust monitoring studies should be directed to the measurement of total dust unless there are other indications that the respirable fraction (notionally 1–7 μm) poses a particular hazard.

The measurement of respirable dust (e.g. by means of a miniature cyclone) has been described by Nutt:[4] in the cases of talc, and precipitated and fumed silica, occupational exposure limits for both total and respirable dust exist.[2,3]

3.3. Local Exhaust Ventilation

The provision of ventilation is the engineering approach to the control of airborne contamination. Applications are widespread and need not really be concerned with reducing pollution, since the suppression of explosion risks and the control of heat and humidity are also major application areas. Local exhaust ventilation (LEV) is the means of capturing contaminants at or near their source and transporting them (e.g. via a ducted system) for discharge at a safe point of disposal.[35] Implicit in this concept is the consideration that the design and operation of LEV systems must be specific to 'local' requirements, but outline comment is appropriate here in view of the recognized deficiency[36] of many installations.

3.3.1. Velocity Profiles

The likelihood of a particle being deflected by an air stream will be dependent on inertial factors already discussed. From the outset it must be clear that the deflection of high inertia particles in this way cannot be easy. Indeed the ability to generate a rapid and directional air flow around an orifice which is exhausting is not easy either. In essence the orifice can be regarded as an evacuated hole, into which air can pour from all directions under a pressure gradient which is at best one atmosphere. This is a markedly different situation from that where the orifice is driving out air (i.e. blowing) and the directional effect can be maintained over a considerable distance beyond the plane of the opening.[37]

Much of what is currently known about velocity profiles around an exhaust opening can be attributed to the work of DallaValle published more than 30 years ago.[38,39] He obtained velocity contours and streamlines for both square and circular openings and showed for each that the axial velocity at a distance equal to the width of the opening was no more than 10% of the average face velocity.[38] Air was shown to be drawn in from all directions including from behind for a plain orifice. A simple flange on the orifice can eliminate the flow from behind enabling reductions in total flow of about one-third for the same axial velocities in front of the opening. DallaValle also enunciated a similarity principle which states that velocity contours, when expressed as a proportion of the face velocity at the opening, are functions only of the shape of the hood. Thus the contours are identical for similar hood shapes when reduced to the same basis of comparison.

3.3.2. More Contemporary Studies

The problem of obtaining air velocities from other than experimental measurements has provided the basis of more contemporary studies.

What relationships have been developed have been largely empirical.[39] One simple expression is given in eqn (10), which gives the centreline velocity (v) at a distance x from a circular or square duct of area A, where Q is the volume rate of air flow.[38]

$$v = Q/(10x^2 + A) \tag{10}$$

Conventionally these empirical relationships have only provided centreline velocities, but a recent innovation has been the derivation by Flynn and Ellenbecker of a theoretical equation (eqn (11)) which allows calculation of the velocity at any point in front of a circular flanged hood.[39] This expression is given in terms of the radius (a) of the circular opening and a parameter ϵ which defines the position of the point in circular coordinates.

$$v = \sqrt{3}\, Q\epsilon^2/(2\pi a^3\sqrt{3 - 2\epsilon^2}) \tag{11}$$

Both these equations can give velocities in relation to the mean face velocity ($V = Q/\pi a^2$ for a circular section duct) and calculated values of v/V for centreline flow are given in Fig. 7 alongside experimental data.

The marked decrease in velocity with even small distances from the exhaust opening is the basis for the recommendation that local exhaust hoods must not be contemplated for any process which cannot be conducted in the immediate vicinity of the hood.[36]

3.3.3. Effective Design

It follows from all these considerations that an LEV system must be carefully designed if it is to provide worthwhile gains with respect to airborne dust control. For example, the hood may be extended to include side screens to enhance directional effects.[40] Performance limitations should be recognized, and if it is unlikely that the inertial effects of a particle can be effectively countered, then such inertia should be exploited to appropriate advantage. Where particles are being generated with a preferential direction of motion (e.g. tangentially from a grinding wheel), then the exhaust opening should be orientated to intercept these without unnecessary deflections. This means that for particles being discharged downwards (e.g. bag-emptying) the preferential direction for exhausting is also downwards.

Despite limitations, which may be with respect to both performance and cost, local exhaust ventilation remains a realistic option: LEV

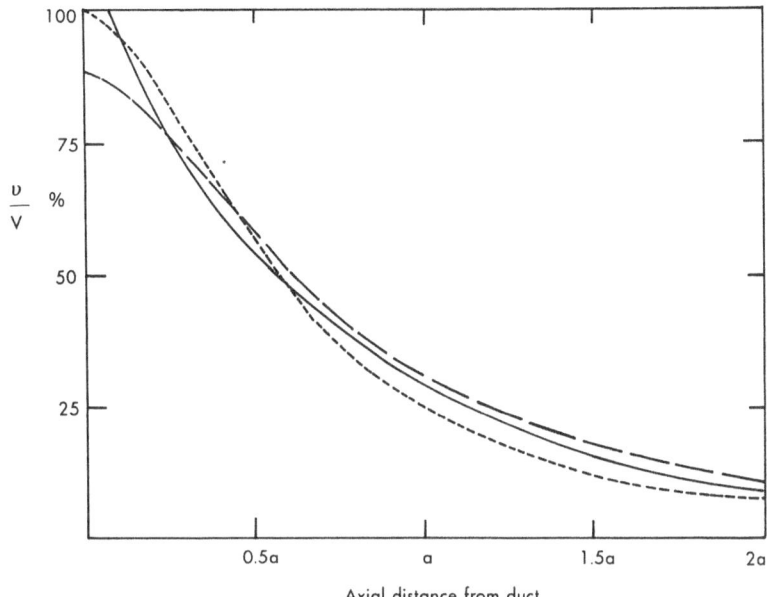

Fig. 7. Reduction of axial velocity with increasing distance from a circular exhaust duct of radius a. The axial velocity is shown as a percentage of the mean face velocity V, and the distance from the duct as a multiple of the radius a. The three traces shown are calculated values derived from eqns (10) (- - - -) and (11) (- -) alongside experimental data[39] for a flanged hood (—).

systems are in widespread use and the subject of continued development and study.[39,41,42] Indeed, there still remain many situations where LEV offers the only viable approach, a point which emphasizes the need for effective design. Of course, achieving an effective installation means considering more than just the hood region and examples of preferred practice are given in a manual[36] and a handbook.[40] The subject has also been reviewed and areas for priority attention summarized in terms of nine basic design factors:[43,44]

(1) minimizing capture distance;
(2) providing adequate capture velocity (or adequate exhaust rate);
(3) using gravitational force;
(4) providing adequate conveying velocity for particulates;
(5) minimizing pressure drop;

(6) incorporating containment wherever possible;
(7) providing even airflow distribution at hood faces;
(8) placing the fan as the last system element;
(9) installing a static pressure gauge and testing the system.

3.4. Control at Source

Control at source is a term often applied to any localized containment (including LEV) but within the context of rubber processing can be applied more specifically to mean controlling the source of the dust. This has been made possible for operations with many rubber chemicals by replacing traditional powders with less dusty forms. This may be achieved even with the minimal addition of coagents (e.g. oil) to give either dust-suppressed (DS) powders or, after compaction, granules, tablets or pastilles.[22,45] Spivey has studied the dust-generating potential of alternative forms in terms of the airborne dust generated above a hopper or beneath a funnel, measuring, respectively, airborne concentrations and a photometric dust index.[22] Table 3 lists some results for four accelerators.

Although significant improvements can be achieved it should be noted that dust is not completely suppressed and pelletized forms may become dusty again by attrition in handling. Products in these forms are prepared with no more than 5% of suppressant additive and further improvements would be expected with larger additions. Grades

TABLE 3
DUST-GENERATING POTENTIAL OF ALTERNATIVE FORMS
OF RUBBER CHEMICALS

	Sample point height (cm)	Total dust (mg/m³)	Dust index
MBT normal powder	30	22·2	300
MBT DS powder	30	3·3	40
MBTS normal powder	30	49	220
MBTS DS powder	30	17	40
MBTS normal powder	10	149	220
MBTS DS powder	10	37	40
NOBS 7 mm pellet	10	10·0	0
NOBS millipellet	10	2·2	0
DPG 4 mm pellet	—	—	10
DPG 2 mm pellet	10	2·0	0

containing up to 50% of such additives are called 'predispersed' and typically employ waxes or polymers as the dispersion medium.[22,46,47] Such grades may vary in form from pastes to extruded rods and when impregnated with a polymer (e.g. with 20% EPDM) are often called 'polymer-bound'. The term 'polymer-bound' applies too to systems where the active component is incorporated as a co-monomer in the base polymer,[48] an approach which also merits inclusion here.

Undoubtedly there is considerable scope for environmental benefit amongst the options of physical type now available, although more quantitative comparisons of the extent of such benefits would be valuable. The wide range of choice available was covered in one review when some 15 different delivery forms for rubber chemicals were identified.[49]

In assessing these developments it should be recognized from the earlier discussions that dust when airborne in the workplace is not readily amenable to containment measures and, furthermore, that engineering approaches are not cheap.[50,51] The selection of materials to ensure that no dust is generated must be regarded as conceptually the simplest and most effective method for dust control.

4. VAPOURS

If should be made clear from the outset that, with very few exceptions, vapours are invisible. Whether a material in the vapour phase can be visibly revealed by its condensation into a cloud of droplets depends on particular circumstances, and the absence of a mist cannot guarantee the absence of vapour. Thus what can be seen is no guide at all to the existence, or otherwise, of a vapour hazard.

4.1. Mists and Vapours

Chemically mists and vapours must be similar, but physically their characteristics are markedly different. Airborne droplets if stable (i.e. if not liable to re-evaporation) can be considered physically as an additional pollutant in a condensed phase, to behave as discussed earlier for dusts. Indeed the generic name 'aerosol' refers to any gasborne dispersion of small particles irrespective of whether those particles are liquid or solid.[33] It is the total aerosol burden which is determined by sampling through a filter, and the cyclohexane-soluble (i.e. notionally organic) portion of that burden which forms the basis of one standard method of quantifying airborne fume in the industry.[3] The diversity of species amenable to capture by this technique is the basis for the recommendation that limits intended for the visible portion of hot rubber fume should not be applied to areas substantially contaminated by organic dusts. For the record, the recommended limits on the aerosol fraction of hot rubber fume are:

Personal monitoring: 0.75 mg/m^3 (8h TWA, reducing to 0.6 mg/m^3 in January, 1990).
Static monitoring: 0.25 mg/m^3 (8h TWA).

Although physically identifiable with the airborne particulates this condensable material has its origins in the vapour phase, a factor which should be borne in mind in any consideration of the chemical nature of workplace vapours. That attention to the nature of such vapours is important to the rubber industry is evident not only from a history of health problems,[4] but also because the very nature of these vapours is not readily predicted. With the exception of solvent-based operations (e.g. degreasing, bonding, etc.), the industry attempts to avoid using notably volatile materials in its formulations or processes. On this basis a consideration of the known ingredients of rubber formulations would appear to offer little direct guidance. Coupled with the difficulty of even seeing where material is being volatilized, the recognition, identification and quantification of vapour hazards emerge as tasks of some difficulty. Not surprisingly this subject has become the focus of much attention and research.

4.2. Generation and Dispersal of Vapours
Firstly it may be beneficial to reflect on the physical nature of gases and vapours. Here molecules are no longer associated with one another and must be treated individually.

4.2.1. Vapour Velocity
An isolated molecule of gas has a velocity which is dependent on the Kelvin temperature (T) according to the classical equation (eqn (12)) which gives the root mean square speed C in cm/s when the gas constant R is conventionally expressed in cgs units.

$$C = \sqrt{3RT/M} \tag{12}$$

On this basis a molecule of carbon disulphide (mol. wt $M = 76$) has an RMS speed of nearly 400 m/s at 200°C.

Such exceptionally high speeds should not be regarded as indicative of the speeds by which gases disperse, as the molecules themselves undergo frequent collisions (e.g. with air molecules also travelling at similar speeds even at ambient temperature), covering perhaps only a few nanometres between successive changes in direction.[52] Nevertheless, molecular motion provides the mechanism for dispersal and undoubtedly, for gaseous mixing, rapid and effective processes operate. Engineering the containment of this disposal process by local exhaust ventilation can be viewed as a local dilution approach where the most contaminated air is continually being replaced by less contaminated air.[53] On this basis the suitability and design of an LEV system for vapour emission control becomes critically dependent on the toxicity of that vapour and its rate of emissions into the air.[36,53]

4.2.2. Vapour Generation from a Liquid

The generation of vapour from a condensed phase is also a temperature-dependent effect, the simple case of the equilibrium vapour pressure (P) above a liquid being determined by the Clausius–Clapeyron equation (eqn (13)).[54]

$$P = P_0 \exp(-L/RT) \qquad (13)$$

In this equation L is the molar latent heat of evaporation and P_0 is the pre-exponential constant. The exponential increase of vapour pressure with increasing temperature has direct relevance to the evaporation of solvents and the effect can be illustrated by comparing the vapour pressures at a single temperature for solvents having different boiling points (Table 4). The boiling point is simply the temperature when the vapour pressure equals the external pressure. Alongside these results are also tabulated values of the respective occupational exposure limits in units (parts per million by volume) which effectively reflect component partial pressures. Popendorf[53] has proposed that both volatility and exposure limit be combined into a single factor to facilitate comparisons amongst different materials. For example, he defines a Vapour Hazard Index (VHI) according to eqn (14) and obtains values from 4·1 (benzene) to 1·5 (nonane) for the chemicals listed in Table 4.

$$VHI = \log_{10}\left(\frac{\text{vapour pressure, ppm}}{\text{allowable limit, ppm}}\right) \qquad (14)$$

TABLE 4
VOLATILITY DATA AND OCCUPATIONAL EXPOSURE LIMITS FOR
SELECTED INDUSTRIAL SOLVENTS[2,53]

Material	BP (°C)	P at 25°C (mm Hg)	OEL 8h TWA[a] (ppm)
Dichloromethane	40	431	100 (C)[c]
Acetone	56	230	1000
Methanol	65	122	200
Hexane	69	151	100
1,1,1-Trichloroethane	74	121	350 (C)[c]
Benzene	80	95	10
Trichloroethylene	87	74	100 (C)[c]
Toluene	111	28	100
1,1,2-Trichloroethane	114	24	10
Methyl isobutyl ketone	117	19	100
o-Xylene	144	6·6	100
Nonane	151	4·3	[b]

[a] Time weighted average values.
[b] 200 ppm, ACGIH.
[c] (C), Control limit.

4.2.3. Generation from Rubber

The generation of a vapour from a condensed phase which is a rubber
is an inherently more complex situation than for a single-component
solvent. There may be no reason to suppose that the actual threshold
for volatilization of a component will change, no change being found
in the case of plasticizer evolution from vulcanized NBR,[55] but rates of
volatilization may be different. For example Luston[56] describes the
physical loss of stabilizers from polymers as a process which appears to
be diffusion-controlled. However, like the volatilization of a liquid,
diffusion is also considered to be a thermally activated process; the
coefficient of diffusion D bears an exponential dependence on temperature in the same manner as given in eqn (13).[57] In the relationship
for D given below (eqn (15)), D_0 is the pre-exponential factor and E_D
the activation energy for diffusion.

$$D = D_0 \exp\left(-E_D/RT\right) \qquad (15)$$

The relative contributions of diffusion and vaporization have not been
critically examined for volatile emissions from rubber, although the

temperature dependence of such emissions has been clearly demonstrated. For example Willoughby has observed a four-fold increase in weight loss from a single-curing formulation when the curing temperature is increased from 160°C to 190°C, and has reported the decay of vapour emissions as a freshly vulcanized rubber cools from the press (Fig. 8).[50] Whereas the latter measurement was achieved using a continuously monitoring instrument, the decay of vapour emissions from cooling rubber has also been demonstrated by Berg, Olsen and Pedesen who collected separately the vapours evolved during different intervals from the onset of cooling.[58,59] Table 5 gives the total vapour yields calculated from the results reported for three intervals covering the first 45 min after cure. Differences in the rates of vapour release reflect differences in the chemical compositions of these vapour mixtures.

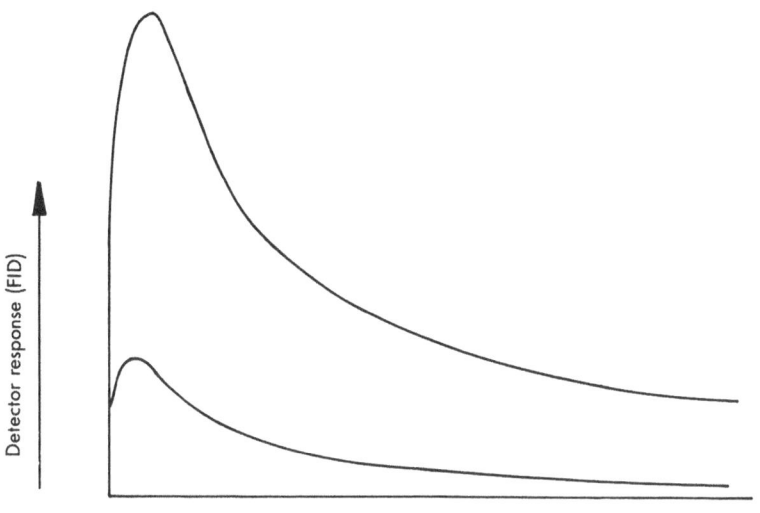

FIG. 8. Decay of emissions from a cooling vulcanizate. Traces for two different compounds are shown, the formulations used being the same except for the type of base polymer used, either an emulsion (upper trace) or solution (lower trace) SBR. Arrangements for cure and monitoring were the same in both cases, the vapour emissions being monitored during the first 10 min of cooling from a 200°C press cure. A portable flame ionization detector, capable of responding to a wide range of organic vapours, provided the quantitative comparison. (Reproduced by courtesy of the PRI, UK.)

TABLE 5

TOTAL YIELDS OF VAPOURS FROM SELECTED RUBBER VULCANISATES DURING
15-MINUTE INTERVALS FROM THE ONSET OF COOLING

Formulation[a]	Total of all vapours collected (mg/kg rubber)	Vapour yields, 0–45 min Percentage distribution		
		0–15 min	15–30 min	30–45 min
A	130	63	25	12
B	210	78	18	4
C	110	79	18	3
D	510	96	4	<1

[a] A, accelerated S cure of EPDM at 180°C. B, accelerated S cure of NBR at 175°C. C, accelerated S cure of CR/BR at 188°C. D, peroxide cure of EPDM at 170°C.

The chemical composition of such vapour mixes will be discussed subsequently (Section 5.3, etc.), but the existence of residual emissions after a substantial period of cooling is a consequence of the ability of chemical operations (i.e. vulcanization) to generate species more volatile than the original ingredients of the rubber compound. Thus heat should be recognized as the driving force for the immediate release of vapours, but heat can also be a factor in the initiation of some long-term vapour emissions. Thus *hot* rubber and *heated* rubber may be the source of vapour hazards. Exposure to fume is believed to be a significant factor in the increased incidence of lung cancer and other diseases, but curing-room operatives would not be the only personnel at risk.[3,4,23]

4.3. Vapour Mixtures and Their Analysis
Before considering further the chemical mixture of species likely to be released from rubber it will be valuable to consider briefly how such identifications can be made and to note conventions in the expression of component concentrations.

4.3.1. Measures of Concentration
At least two measures of concentration for airborne gases and vapours are in common use and these can be interconverted by reference to the universal gas law which applies to either total or partial pressures.

Equation (16) gives the expression for the partial pressure P_i of a component i in terms of the number of moles n_i present in a total volume V (R and T as defined previously).[52,53]

$$P_i V = n_i R T \tag{16}$$

The respective concentrations in molar (n_i/n) terms and weight per volume (C_i) are obtained from eqns (17) and (18) respectively:

$$\frac{n_i}{n} = \frac{P_i}{P} \tag{17}$$

$$C_i = \frac{n_i M_i}{V} \tag{18}$$

where n is the total number of moles and P is the total pressure. When P_i is in atmospheres and R in units of litre atm/K mol, then the conversion between fractional molar concentration and C_i in g/litre is given by eqn (19).

$$\frac{n_i}{n} = \frac{RT}{M_i} \cdot C_i \tag{19}$$

If the C_i term has units of mg/m^3, then n_i/n is obtained not as a fractional concentration but in parts per million (ppm). When T is 298 K (25°C) the conversion factor becomes:

$$1 \text{ ppm} = \frac{M_i}{24 \cdot 5} \text{ mg/m}^3$$

These are the units of airborne concentration in common use. For a mixture of different components, each will have a specific conversion factor (dependent on M_i), and the two conventions may produce different rankings in component concentrations. For example, the reaction of eqn (20) has the capacity to generate equimolar concentrations of dodecene ($M = 168$) and hydrogen sulphide ($M = 34 \cdot 1$); yet 1 ppm of each would be equivalent to 6·9 and 1·4 mg/m^3 respectively. It is evident from eqn (16) that ppm in molar terms are also ppm in terms of pressure or volume (when 1 ppm is conceptually equivalent to 1 ml of component i at atmospheric pressure dispersed in 1 m^3 of air). Whereas these three terms are interchangeable with one another, they are not interchangeable with ppm in weight terms. This point is made to obviate a potential confusion which may arise from the frequent use of ppm by weight to describe trace component concentrations in a condensed phase (e.g. monomer in polymer).

4.3.2. Contaminant Monitoring

It is not proposed to discuss the various techniques by which airborne concentrations of contaminants can be determined. These are analytical details and the subject of prolific publications, but one outline scheme[60] merits inclusion (Fig. 9) to illustrate the range of options available. A point worthy of comment is the increased interest in

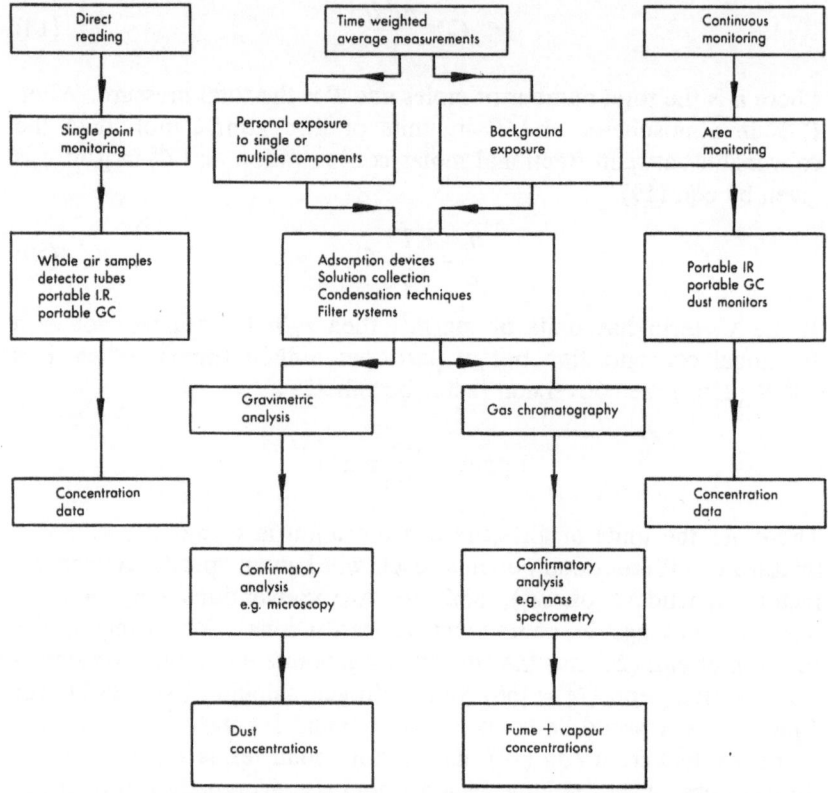

Fig. 9. Summary of on-site monitoring techniques. In this format the range of available options is broken down into three groups: instantaneous, time weighted average and continuous monitoring techniques. (Reproduced by courtesy of the Society of Environmental Engineers, UK.)

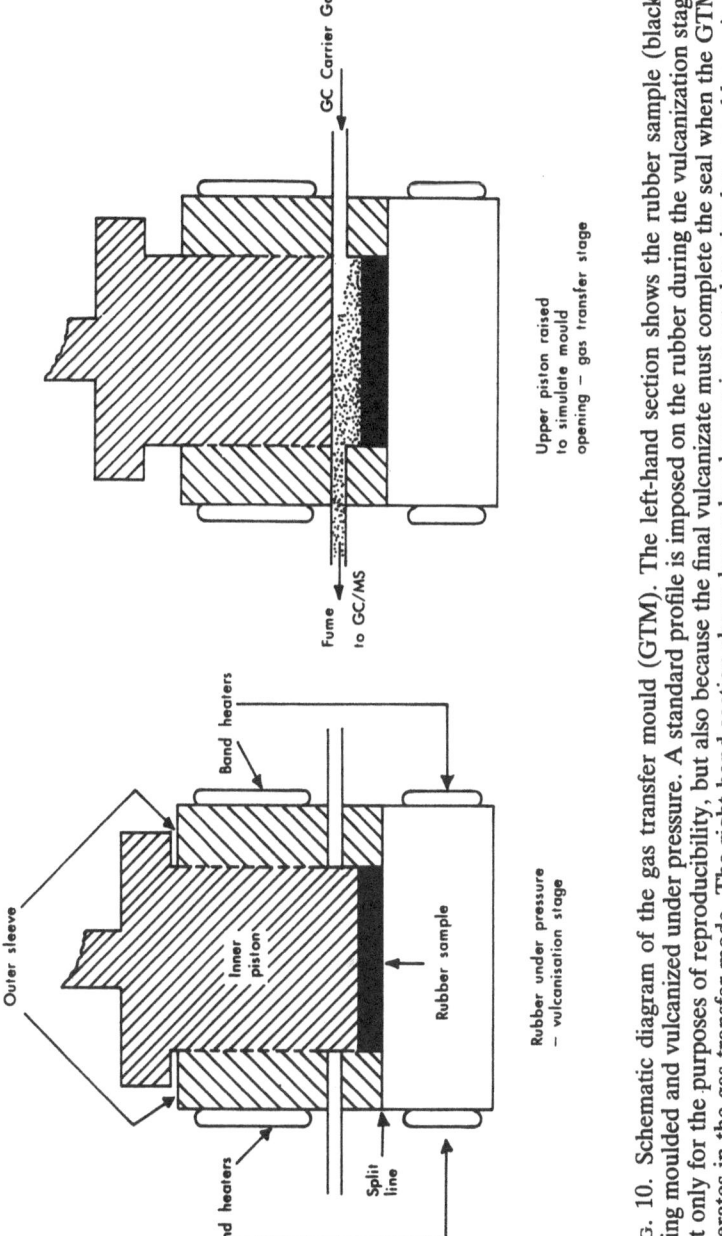

FIG. 10. Schematic diagram of the gas transfer mould (GTM). The left-hand section shows the rubber sample (black) being moulded and vulcanized under pressure. A standard profile is imposed on the rubber during the vulcanization stage not only for the purposes of reproducibility, but also because the final vulcanizate must complete the seal when the GTM operates in the gas transfer mode. The right-hand section shows how a head space is created to simulate mould opening, and through which a gas flow passes to effect the vapour transfer. (Reproduced by courtesy of Rapra Technology Ltd, UK.)

adsorption techniques for sample collection prior to analysis: techniques pioneered for monitoring industrial atmospheres less than 20 years ago[61] and already employed in several studies of air in the rubber manufacturing workplace.[58,59,62,63] The undoubted convenience of the approach has initiated the development of newer so-called 'passive' samplers,[64-66] whilst enabling the pumped charcoal tube to gain such widespread acceptance that its suitability to an application may not be questioned. Yet its suitability to the analysis of mixed organic vapours was the subject of the earliest cautionary comment,[61] with advice both then and since on the likely displacement from charcoal of polar organics by more strongly adsorbing non-polar species.[67]

One problem in assessing the reliability of an atmospheric monitoring technique is the generation of suitable reference measurements. A recent and extensive validation exercise for charcoal tube sampling

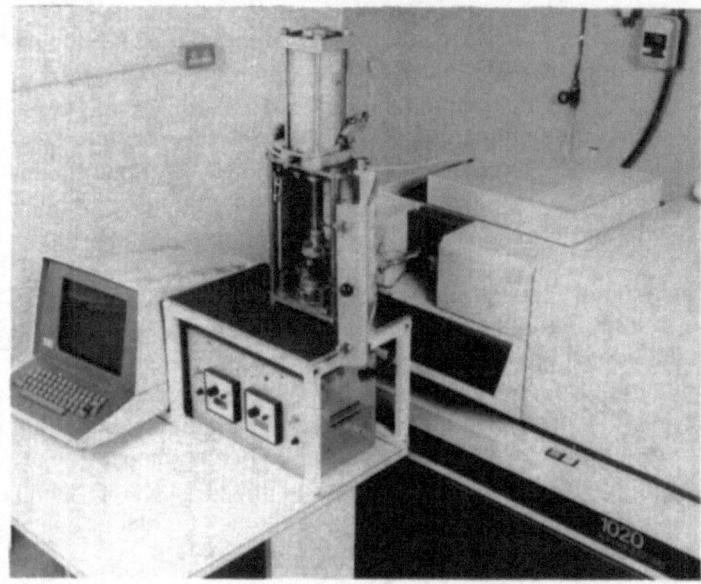

FIG. 11. Direct analysis of vulcanization fume using the gas transfer mould operating on-line a Finnigan 1020 automated GC/MS. The mass spectrometer has the capability to provide diagnostic spectra on component quantities of 1 µg or less. These components are separated for analysis by the gas chromatograph which in this installation is a Perkin–Elmer Sigma 3 and is directly behind the GTM. (Reproduced by courtesy of Rapra Technology Ltd, UK.)

used a dedicated laboratory assembly equipped with on-line gas chromatography,[68] an approach which could be followed in studies of vulcanization fume. Recent developments in technique have enabled vulcanization fume to be analysed directly in an integrated mould (Fig. 10) and gas chromatograph, an instrument equipped in this case with on-line mass spectrometry and computerized data handling (Fig. 11).[69] This laboratory facility has the capability to fractionate and analyse the components of the vapours as they are generated at the instant of mould opening, and should bring a new confidence to this field of analysis.

5. THE NATURE OF WORKPLACE VAPOURS

Experience dictates that air pollution problems are not easily rationalized. Well-documented issues such as photochemical smog[70] or the fate of atmospheric chlorofluorocarbons[71] have revealed the potential for chemistry of considerable complexity, and even it seems for a different result (i.e. ozone enrichment or depletion) from not too dissimilar photochemical reactions. Smoke or acid gases from combustion processes continue to cause concern,[72-74] and informed exchanges highlight the problems of identifying a source even for well-defined emissions.

5.1. Background Air
Such issues emphasize the point that air pollution is not confined to the workplace, and by implication that some components of workplace pollution may have their origins elsewhere. This latter point has a bearing on the monitoring of polycyclic aromatic hydrocarbons in factory air. Whereas polycyclic aromatic hydrocarbons might be expected in emissions from high aromatic process oils,[4] they are also common[72] air pollutants. Table 6 shows the results obtained by Willoughby for benzo(a)pyrene concentrations in the *outside* air at Shawbury, results which show a marked variation dependent on meteorological factors.[75] This variation reflected that being observed for measurements of benzo(a)pyrene concentrations inside rubber factories, and Nutt[4,76] carried out simultaneous monitoring of workplace and outside air and found no significant excess of benzo(a)pyrene in the factory.

Considerations of combustion processes as a source of air pollution

TABLE 6
BENZO(a)PYRENE IN THE OUTSIDE AIR AT
SHAWBURY DURING SUMMER AND AUTUMN 1974

Sampling date	Sampling times	Benzo(a)pyrene (ng/m^3)
18 June	9.00–17.00	Detected, but <1
18–19 June	21.00– 5.00	Detected, but <1
19 June	9.00–17.00	None detected
19–20 June	21.00– 5.00	2·4
20 June	9.00–17.00	None detected
5 November	4.00–12.00	17
6 November	4.00–12.00	14
7 November	4.00–12.00	3·9
8 November	4.00–12.00	1·2
9 November	4.00–12.00	1·2
10 November	4.00–12.00	None detected

should also take account of in-house combustion, as for example with the operation of fork-lift trucks. The general pattern of exhaust emissions from motor vehicles are well documented, an 'untreated' vehicle giving rise to up to 5% carbon monoxide and up to 3000 ppm of nitrogen oxides in typical exhaust gas compositions.[77] Carbon monoxide emissions are dependent on the efficiency of combustion (e.g. fuel–air ratio) and nitrogen oxides on temperatures within the combustion chamber. In any event both are dependent on the mode of operation of the engine and results for an LP gas engine show the highest levels of carbon monoxide (e.g. up to around 2·5% or 25 000 ppm) are obtained at low engine speeds on full load.[78] More than 90% of this can be removed by use of a platinum-based catalyst operating at elevated temperature within the exhaust system. However, the catalytic oxidation of exhaust gases is ineffective for the removal of nitrogen oxides, an alternative approach here being the use of exhaust gas recycling (to cool the flame front) which has been claimed to reduce nitric oxide emissions by as much as 60%.[77] All these gases possess recognized toxicity (occupational exposure limits:[2] CO, 50 ppm; NO, 25 ppm; NO_2, 1 ppm), and clearly the safe operation of internal combustion engines within the workplace merits particular attention.

It will be clear from these examples that the emerging knowledge of external air pollution can have a direct relevance to the situation in the workplace. The chemistry is potentially complex, and the more species that are involved then the greater is the opportunity for chemical change. Airborne vapours may be species generated locally or from a remote source: it is not unknown for prevailing or enforced air currents to deliver to one process area the emissions from another.

5.2. Vulcanization Fume

Investigations of the species to be found in the vapours and mists from vulcanization have been mainly confined to the last ten years, and the current understanding can be attributed largely to laboratory studies. An essential ingredient of much of this work has been gas chromatography with on-line mass spectrometry (GC/MS) and the development of vulcanization techniques which can most effectively exploit this instrumentation has been described and reviewed.[69,79] The gas transfer mould already described (Figs 10 and 11) represents the latest stage in the evolution of these techniques.

5.2.1. Contributions from the Base Polymer

A recognizable source of workplace air pollution is the base polymer and specifically the volatile species within it.

5.2.1.1. Low molecular weight impurities. Even trace quantities of low molecular weight impurities within the polymer can make a significant contribution to process vapours when the polymer itself is characteristically such an abundant component of the rubber compound. An impression of the types of species which may be present can be gained by considering the outline scheme in Fig. 12. Whether such species remain in significant quantity in the product polymer depends on the efficiency of the stripping stage and it may be that there are practical difficulties here with respect to latices or for especially tacky polymers.[79]

The monitoring of airborne monomer concentrations around subsequent processing operations may provide one index of the efficiency of such stripping. For example, Nutt[76] has found chloroprene concentrations as high as 20 ppm around polychloroprene latex dipping, whilst 18–45 ppm of methyl acrylate has been found next to dies during the extrusion of 'Vamac' ethylene/acrylic rubber at temperatures between 71 and 93°C.[80]

5.2.1.2. Dimers and trimers. With the exceptions of isoprene and the low molecular weight aliphatic olefins (ethylene, propylene,

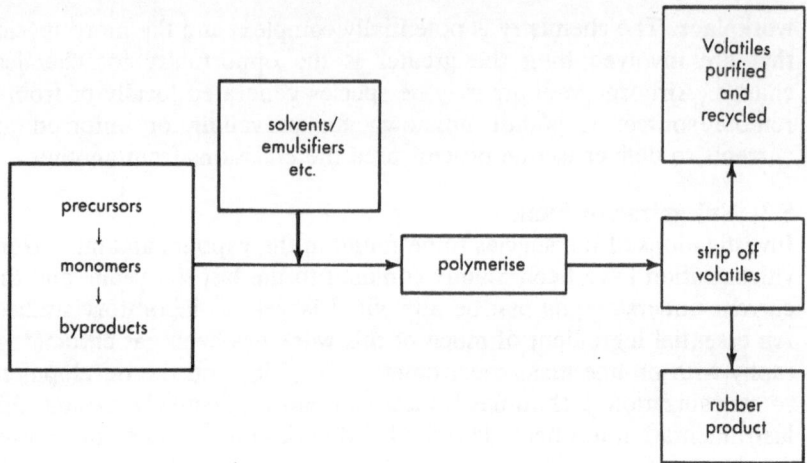

FIG. 12. Outline scheme for base polymer manufacture.

isobutylene), the monomers for the most common rubbers have designated occupation exposure limits, or threshold limit values, ranging from 1000 ppm for butadiene to 2 ppm (control limit) for acrylonitrile.[2,3] However, it will be evident from the scheme in Fig. 12 that monomers are not the only species likely to be encountered as the residues of polymer manufacture. Dimers and trimers may be found although their recognition as such may not be clear from the applied nomenclature (see also Table 7).[79] In the case of silicone rubbers, cyclic siloxanes containing up to as many as 16 repeating units have been identified amongst the vapours of vulcanization.[59] For factory air studies, trimers of both butadiene[62] and isobutylene[63] have been detected.

The trimer of isobutylene merits special mention, for it has been found to be one of the most abundant species in vulcanization fume from elastomers not derived from isobutylene as a monomer. This effect was first explained in 1981 from the results of laboratory vulcanizations at Rapra.[81] This isobutylene trimer is not a single chemical compound, but is a collection of isomeric C_{12} olefins (dodecenes) and is obtained when an appropriate mixture of isomeric dodecyl mercaptans is heated in a vulcanization mix. Such a mixture of mercaptans is a commonly used ingredient of an emulsion polymerization, such as for SBR or NBR, and the interconversion between isomeric mercaptans and olefins is illustrated by the representative examples given in eqn (20).[81,82] This outline scheme shows some of the principal isomers expected when the starting material is isobutylene.

$$C_9H_{19}-\underset{\underset{Me}{|}}{C}=CH_2$$

$$\underset{-H_2S}{\overset{H_2S}{\rightleftharpoons}} \quad C_9H_{19}-\underset{\underset{Me}{|}}{\overset{\overset{Me}{|}}{C}}-SH$$

$$C_8H_{17}-CH=CMe_2$$

$$3Me_2C=CH_2 \longrightarrow$$
isobutylene

$$(C_5H_{11})_2C=CH_2$$

$$\underset{-H_2S}{\overset{H_2S}{\rightleftharpoons}} \quad (C_5H_{11})_2\underset{\underset{Me}{|}}{C}-SH$$

$$C_4H_9-CH=\underset{\underset{Me}{|}}{C}-C_5H_{11}$$

mixture of
dodecyl mercaptans

mixture of
dodecenes

(20)

The important contribution of residues of the polymerization to the type and quantity of subsequent vapour emissions can be demonstrated, perhaps to surprising effect, by reference to Fig. 8. The traces shown were for two analyses conducted under identical conditions but using vulcanizates which differed only with respect to the *grade* of base polymer used. The upper trace was obtained from a formulation based on an emulsion SBR, and the lower trace correspondingly from a solution SBR mix.

5.2.1.3. Residual components. Residual components from monomer manufacturing may be numerous especially when an energetic process is involved, as is the case for styrene manufacture (eqns (21)–(23)).[83]

$$\overset{600-650°C}{\underset{\text{oxide catalysts}}{\longrightarrow}} \quad C_6H_5CH=CH_2 + H_2 \quad (21)$$

$$C_6H_5CH_2CH_3 \qquad C_6H_6 + CH_3CH_3 \quad (22)$$

$$\underset{H_2}{} \qquad C_6H_5CH_3 + CH_4 \quad (23)$$

Other by-products are obtained in addition to benzene (eqn (22)) and toluene (eqn (23)). Isopropylbenzene (cumene), α-methylstyrene and the derivatives of diethylbenzene (impurity in ethylbenzene feedstock) are obtained and the total mixture reflects the character of aromatic species in emissions known to have their origins in styrene polymers.[83,84] A similar mix of aromatics is obtained in the vapours from SBR-containing formulations whether or not a process oil is present: the results given in Table 7 compare quantitative data from factory air analysis (charcoal tube sampling)[63] and from a laboratory vulcanization study,[85] and suggest a temperature threshold for major emissions of dodecene.

5.2.1.4. Wide vapour range. It will be clear from these discussions that hydrocarbon polymers have the capacity to generate a wide range of hydrocarbon vapours, vapours which are readily amenable to detection and analysis. A number of these species must be present before heating, to be released during processing or vulcanization, and the most volatile of these will be available for release early in the heating cycle. Benzene in particular has a low occupational exposure limit and a marked volatility (see Table 4) and significant emissions from extrusion formulations (e.g. Table 7) should not be unexpected. However, low-boiling aromatics may also be encountered in solvents[3] and adhesive formulations,[87] and an effective assessment of related hazards should not exclude consideration of alternative sources of these vapours.

5.3. Contributions from the Curing System

Whereas laboratory-based studies have proved valuable for identifying the sources of those emissions readily detectable in factory air (e.g. Table 7), such laboratory studies may be a prerequisite to the recognition of the presence of other vapour components. For example, the detection of polar or highly volatile species is well within the capabilities of laboratory-based studies of vulcanization fume, and polar components as volatile as hydrogen sulphide (BP −61°C) or as involatile as aniline (BP 184°C) may be handled using standard procedures and without prior recognition of their presence.[86] In one such study C_1–C_{34} organics were identified in a single vapour mix. It is by recourse to laboratory vulcanizations that some of the by-products of reactions in the rubber can be recognized, and work in this field has been recently reviewed.[79]

TABLE 7

SOME MAJOR COMPONENT HYDROCARBONS FOUND IN EMISSIONS FROM FORMULATIONS CONTAINING SBR

Emission	Hydrocarbon content[a]			
	Extrusion[b] up to 110°C ($\mu g/m^3$)	Vulcanization[c] at 145–165°C ($\mu g/m^3$)	Vulcanization[c] at 180–240°C ($\mu g/m^3$)	Vulcanization[d] at 200°C ($\mu g/50g$ of compound)
Benzene	25–180	10–1200[e]	8–15	190
Toluene	20–160	6–800[e]	4–8	D[f]
Styrene	1–20	2–180	90–500	570
Ethylbenzene	1–15	2–90	30–150	D
4-Vinylcyclohexene[g]	0–3	ND	30–210	D
Isopropylbenzene	0–10	2–200	60–250	220
Di-isopropylbenzene	0–7	1–75	35–70	450
Dodecenes	0–20	5–180	300–7000	12 600
Cyclododecatriene[h]	0–10	5–400	ND	D

[a] Abbreviations: D, detected but not quantified; ND, not detected.
[b] NR/SBR/BR tyre-tread compound.
[c] NR/SBR shoe-soling compound.
[d] Reference SBR compound containing no oil.
[e] A high result which may indicate contributions from another source.
[f] Detected in associated studies[86] employing different GC conditions.
[g] Dimer of butadiene.
[h] Trimer of butadiene.

5.3.1. Generation of New Vapour Products

One feature which is fundamental to the current understanding is that the chemical changes during vulcanization do generate new vapour products, many having substantial volatility, a point already referred to in Section 4.2. If trends with regard to specific emissions from particular ingredients can be recognized, then it is probably the delayed-action curing systems which currently offer scope for the most coherent rationalization at present. In these cases the active curing agent is only formed after a chemical reaction has released some ancillary fragment from the parent molecule. Some examples are given in eqns (24)–(26) for both sulphur and non-sulphur chemicals. The release of amines from sulphenamide cures has been observed in several studies,[58,86,88] that of phthalimide from PVI has been reported,[3] whilst other work has found cinnamaldehyde from a fluoroelastomer cure.[89]

$$\text{(benzothiazole)} \ C\text{--}S\text{--}NR_2 \longrightarrow \text{(benzothiazole)} \ C\text{--}S\text{--}SX + R_2NH \quad (24)$$
$$\text{amine}$$

$$\text{cyclohexyl}\text{--}S\text{--}N(\text{phthalimide}) \xrightarrow{\text{HSX}} \text{cyclohexyl}\text{--}S\text{--}SX + \text{(phthalimide)}NH \quad (25)$$
$$\text{phthalimide}$$

$$(CH_2)_6 \begin{array}{c} N\text{=}CHCH\text{=}CHPh \\ \\ N\text{=}CHCH\text{=}CHPh \end{array} \xrightarrow{H_2O} (CH_2) \begin{array}{c} NH_2 \\ \\ NH_2 \end{array} + PhCH\text{=}CHCHO \quad (26)$$
$$\text{cinnamaldehyde}$$

The volatile by-products of subsequent vulcanization processes are critically dependent on formulation and temperature, and can show variations which are not readily rationalized at present.[79] For example, MBT and its derivatives release benzothiazole,[58,86,90] and not the regenerated MBT which might have been hitherto[91-93] anticipated. Vulcanizations involving thiurams and dithiocarbamates can release amines and carbon disulphide (BP 46°C), via at least two recognized routes (eqns (27)–(28)),[94,95] neither of which would predict the observed release of amine only from TBTD.[85] The efficiency of mixing may be one factor in the release of carbon disulphide,[96] whilst the efficiency[97] of vulcanization may be a factor in the release of hydrogen sulphide.[98] Temperature is certainly a factor in the release of all these vulcanization by-products,[79] hydrogen sulphide emissions increasing markedly above a threshold of around 170°C.[86,98]

$$Me_2NCSSCNMe_2 \xrightarrow{\text{vulc.}} \left[Me_2N\overset{\overset{\displaystyle S}{\|}}{C}SH \right] \longrightarrow Me_2NH + CS_2 \quad (27)$$

$$\xrightarrow{\Delta} Me_2N\overset{\overset{\displaystyle S}{\|}}{C}NMe_2 + S + CS_2 \quad (28)$$

5.3.2. Competing Reactions

With the exception of implications regarding solubility,[96] a feature not yet apparent is any recognized trend in the emissions from sulphur cures with variations in the base polymer. This is not the case with peroxide cures,[99] where the reactivity of the polymer can influence both the quantity and type of emissions. A well-studied example is that of NR which carries an abundance of abstractable allylic hydrogens to favour alcohol formations (eqn (29)). Thus when DTBP (R = Me) is the peroxide, tert-butanol (BP 82°C) is obtained, whilst cumyl alcohol (2-phenyl-2-propanol; BP 202°C) is obtained from Dicup (R = Ph).[100,101] Ketone formation (eqn (30)) competes with hydrogen abstraction and can predominate in the presence of a different polymer:[102] emissions from formulations based on EPDM,[59] silicone and a fluoroelastomer[102] have been characterized. Other by-products include alkenes from alcohol dehydration, although numerous other reactions can occur.

$$R\overset{\overset{\displaystyle Me}{|}}{\underset{\underset{\displaystyle Me}{|}}{C}}O-O-\overset{\overset{\displaystyle Me}{|}}{\underset{\underset{\displaystyle Me}{|}}{C}}R \longrightarrow R\overset{\overset{\displaystyle Me}{|}}{\underset{\underset{\displaystyle Me}{|}}{C}}O\cdot$$

$$\xrightarrow{\text{H}} R\overset{\overset{\displaystyle Me}{|}}{\underset{\underset{\displaystyle Me}{|}}{C}}OH \quad (29)$$

$$\xrightarrow{-Me} \overset{\displaystyle R}{\underset{\displaystyle Me}{}}C=O \quad (30)$$

The existence of an established range of competing reactions is not restricted to peroxide cures. Industrial sulphur cures can bring together such a diverse range of chemicals that there is ample scope for interaction, especially when new components are being generated during cure. For example, the sequence shown in eqns (27) and (28) shows the formation of three products which can all react further under suitable circumstances. Equations (31)–(33) show the involvement of acrylonitrile,[103] a sulphenamide,[104] or nitrosating species[105] in some possible reactions. There is evidence that each of these processes can occur under appropriate conditions in vulcanizing mixes and influence the emissions from vulcanization, an influence which may have profound implications.

$$\xrightarrow{\text{CBS}} \text{MBT} + \text{S} + \left\langle \underset{\hspace{0.5em}}{\bigcirc} \right\rangle\!\!-\!\text{NCS} \qquad (31)$$

cyclohexyl
isothiocyanate

$$\underset{\text{Me}_2\text{N}\overset{\displaystyle\text{S}}{\overset{\|}{\text{C}}}\text{SH}}{} \longrightarrow \text{CS}_2 + \text{Me}_2\text{NH}$$

$$\Big\downarrow \text{CH}_2\!\!=\!\!\text{CHCN}$$

$$\xrightarrow{\text{NOX}} \text{Me}_2\text{NNO} \qquad (32)$$
dimethylnitrosamine

$$\underset{\text{Me}_2\text{N}\overset{\displaystyle\text{S}}{\overset{\|}{\text{C}}}\text{SCH}_2\text{CH}_2\text{CN}}{} \quad (33)$$

5.3.3. Areas of Particular Concern

Emissions of nitrosamines (specifically, N-nitrosamines) arouse particular concern owing to associated cancer risks.[4,106] All secondary amines are potential precursors to N-nitrosamines and it will be apparent from considerations of the reactions both prior to cure and accompanying cure (e.g. eqns (24), (27)) that amines may be commonly encountered in vulcanized mixes and the vapours from them. Some of the amines already encountered in emissions from rubber are listed in Table 8.[79] It should be made clear from the outset that not all the amines listed are secondary, and that primary amines are not direct precursors to N-nitrosamines. Considerable attention has been focused on N-nitrosamine formation within the rubber industry, and various monitoring exercises have detected such nitrosamines in both vapours and solvent extracts.[4,106–109] In one study airborne concentrations in excess of $1\,\text{mg/m}^3$ were found for dimethylnitrosamine (N-nitrosodimethylamine), and as high as $4\cdot6\,\text{mg/m}^3$ for N-nitrosomorpholine.[4] However, the amine is only one-half of the equation and effort has also been directed to establishing the identity

TABLE 8
PRIMARY AND SECONDARY AMINES FROM ACCELERATORS AND CURING AGENTS

Ingredient	Amine evolved		
	Primary	Secondary	BP (°C)
TMTD	—	Dimethylamine	7·4
TBBS	tert-Butylamine	—	44
TETD	—	Diethylamine	56
DIBS	—	Di-isopropylamine	84
NOBS, DTDM	—	Morpholine	128
CBS	Cyclohexylamine	—	135
TBTD	—	Dibutylamine	159
DPG	Aniline	—	184

of potential nitrosating agents. Two of these may be the retarder NDPA and sodium nitrite, in either a salt bath or possibly as a residue from an emulsion polymerization.[4,79,81,85] One response to these findings has been the withdrawal in 1980 of NDPA from the UK market.[3]

Another potential nitrosating agent is in the oxides of nitrogen from fork-lift truck exhausts,[110] illustrating the complexity of interactions which can affect safe working practices within the industry.

Alkyl isothiocyanates have been implicated over eye irritation problems,[4,111] and may be obtained from the action of carbon disulphide on certain sulphenamides (eqn (31))[104] or by the breakdown of some thioureas.[4,79] The former process is apparently particularly facile, operating at room temperature: the formation of cyclohexyl isothiocyanate as one of the most abundant components in the emissions of a CBS/TMTD cure is therefore not unexpected.[58]

5.3.4. Suppression by Reaction
Whereas the above discussions concern materials which in combination may increase the hazard, the reaction shown in eqn (33) may serve to reduce a hazard (acrylonitrile): the involvement of this reaction to influence vulcanization emissions is a most recent discovery.[112] There is a clear precedent for this approach in the use of inorganic bases to suppress the release of hydrogen fluoride from FKM

cures.[113] Magnesium oxide and calcium hydroxide were used for this purpose in Viton formulations characterized for their emissions during both cure and postcure. Clearly attention to chemical interactions can lead not only to the avoidance of high-risk situations, but also to the initiation of improvements in present conditions through the selective reduction of emissions. Control at source is a concept not limited to dust suppression; the production of both vapours and extractable species should also fall within its scope.

5.4. Oxidation and Other Decomposition Products

The reactions associated with vulcanization are not the only ones which can lead to the creation of new volatile species. For example, chemical blowing agents are specifically designed to generate gases within the rubber; eqns (34)–(36) provide outline schemes for such changes.[114,115] Although tetramethylsuccinonitrile is a solid which might not be regarded as particularly volatile, airborne concentrations of up to 3 ppm have been measured around trial production runs.[116]

$$H_2NCON{=}NCONH_2 \longrightarrow N_2 + H_2NCONH_2 + CO \qquad (34)$$
$$\text{ADC} \qquad\qquad\qquad\qquad\qquad \text{carbon} \atop \text{monoxide}$$

$$(35)$$

$$Me_2\underset{\underset{CN}{|}}{C}{-}N{=}N{-}\underset{\underset{CN}{|}}{C}Me_2 \longrightarrow N_2 + Me_2\underset{\underset{CN}{|}}{C}{-}\underset{\underset{CN}{|}}{C}Me_2 \qquad (36)$$

$$\text{AZDN} \qquad\qquad \text{tetramethylsuccinonitrile}$$

Other components of a rubber mix which are intended to be involved in chemical changes are the antidegradants. One surprising find of recent studies has been the formation of ketones from certain phenylenediamines apparently by side-chain oxidation.[69,96] In this way IPPD and 6PPD release acetone and MIBK, respectively (eqn (37)), whilst a dialkylphenylenediamine, 77PD, was found to be a particularly prolific source of a methyl amyl ketone presumed to be the isomer shown in eqn (38). This type of reaction may have a parallel in sulphenamide chemistry since cyclohexanone has been obtained as a major component in the emissions from a CBS cure.[58]

$$R-\underset{\underset{Me}{|}}{\overset{\overset{H}{|}}{C}}-\underset{\overset{H}{|}}{N}-\!\!\bigcirc\!\!-\underset{\overset{H}{|}}{N}-\!\!\bigcirc \longrightarrow R-\underset{\underset{Me}{|}}{C}\!\!=\!\!O \qquad (37)$$

$$Me-\underset{\underset{Me}{|}}{\overset{\overset{H}{|}}{C}H}-C_2H_4-\underset{\underset{Me}{|}}{\overset{\overset{H}{|}}{C}}-\underset{\overset{H}{|}}{N}-\!\!\bigcirc\!\!-\underset{\overset{H}{|}}{N}-\underset{\underset{Me}{|}}{\overset{\overset{H}{|}}{C}}-C_2H_4-\underset{\underset{Me}{|}}{\overset{\overset{H}{|}}{C}H}-Me$$

77PD

$$\xrightarrow[\text{mix at 200°C}]{\text{in vulc.}} Me-\underset{\underset{Me}{|}}{\overset{\overset{H}{|}}{C}H}-C_2H_4-\underset{\underset{Me}{|}}{C}\!\!=\!\!O \qquad (38)$$

5-methylhexan-2-one

In general, oxidation products have so far been the subject of only sparse study. Oven-curing operations should provide the maximum opportunity for oxidation, as was observed for Viton E-60C when the press cure yielded both a phosphine and its oxide but the subsequent postcure yielded predominantly the oxide.[89] When benzyltriphenylphosphonium chloride (BTPPC) is in the precursor ingredient, triphenylphosphine oxide is obtained.[117] Emissions of sulphur dioxide from sulphur-containing formulations might be expected to follow a similar pattern but the limited results so far provide inconclusive confirmation.[79] In fact sulphur dioxide has not yet been recognized as a common air pollutant, and its detection from CR formulations[56,86] may suggest the involvement of specific precursor groups, perhaps even α-chlorosulphides as shown schematically in eqn (39).[118,119]

$$\underset{|}{\overset{|}{C}}\!\!=\!\!CH-CH_2-S-\underset{|}{\overset{|}{C}}-Cl \xrightarrow[\text{(ii) Ramberg–Bäcklund reaction}]{\text{(i) oxidation}} \underset{|}{\overset{|}{C}}\!\!=\!\!CH-CH\!\!=\!\!\underset{|}{\overset{|}{C}} + SO_2$$

$$(39)$$

However, it should be recognized that oxidation processes are far from simple and their rationalization may be beyond the scope of even extensive investigations. Nevertheless, an appreciation of the range of such products likely to be released is important, and studies of the change in emissions when hot rubber is allowed to stand in air offer a useful course of action.[58]

5.5. Component Toxicities and Exposure Limits

The concept of a single exposure limit for a fume mix, irrespective of its composition, applies only to the aerosol fraction of hot rubber fumes. For vapours, individual species vary greatly in their modes of physiological action and their activities, and the recognition of appropriate exposure limits relies on some knowledge of the identities of the species present. The toxic effects of different vapours have been reviewed by Dinman.[32] With regard to the volatile hazards from hot rubber, health risks may be posed by the presence of, for example, irritants, suspected carcinogens or embryotoxins, and a current appreciation of the subject has been presented by Buus.[120,121] Occupational exposure limits apply to many of the species concerned and selected examples are given in Table 9.[2]

TABLE 9

UK OCCUPATIONAL EXPOSURE LIMITS (8-HOUR TWA)[2] FOR SOME INDIVIDUAL COMPONENTS ENCOUNTERED IN PROCESS OR VULCANIZATION FUMES

Substance	OEL (ppm)	Substance	OEL (ppm)
Acrylonitrile	2(C)[b]	Dimethylamine	10
Aniline	2	Formaldehyde	2(C)[b]
Benzene	10	Hydrogen fluoride	3
t-Butanol	100	Hydrogen sulphide	10
Butylamine	5	Methyl acrylate	10
Carbon disulphide	10(C)[b]	Methyl isoamyl ketone	100
Carbon monoxide	50	Morpholine	20
Chloroprene	10	Styrene	100(C)[b]
Cyclohexanone	25	Sulphur dioxide	2
Dibutyl phthalate	[a]	Tetramethylsuccino-nitrile	0·5
Diethylamine	25	Toluene	100
Di-isopropylamine	5	Xylene	100

[a] 5 mg/m³.
[b] (C), control limit.

Clearly from the foregoing discussion, the nature and extent of any hazard will depend on the formulation, its history and physical factors such as temperature or surface area. Thus the species released during processing may be different from those released from the same compound on subsequent vulcanization when the opportunity for chemical changes is enhanced. Considerations of sample history should not exclude the consequences of reheating a vulcanizate (e.g. as in intercomponent bonding) which may already contain the volatile by-products of extensive reaction.

5.6. Control at Source
As in the case of dust emissions (Section 3.4), some localized containment of fume may be effected by exhaust ventilation (LEV), although for vapours the control achieved is by dilution (Section 4). Additionally for vapours there is the problem of knowing where to implement controls when there is no visual evidence of emission. The insight gained by studies of emission processes remains fundamental to the practical and effective control of volatile pollutants by any means. Investigations into the quality and quantity of post-vulcanization vapours occupy one priority area of current research effort.

One benefit of the emerging understanding of vulcanization fume is an appreciation of the potential for controlling the release of a vapour from rubber. This is control at source in the stricter sense as previously applied (Section 3.4) to the control of dust generation from solid additives. Examples of the implementation of this approach to control specific component emissions include the removal of Nonox S[122] or NDPA[3] from rubber formulations and their replacement by safer alternatives. Such control at source has been described in general terms as achievable in two ways.[50,79]

(1) Compounding to control by-product volatility.
(2) Modifications to limit the presence or creation of a hazard within the rubber.

The first of these may be viewed as a sort of 'half-way house' where toxic by-products may still be formed but where their emissions are controlled. For example, changes in the type of accelerator will modify the type of amine produced. Thus if the concern is with a fume hazard, a less volatile amine might be generated (e.g. dibutylamine instead of dimethylamine). However, if the major pollution problems are expected to be associated with the rubber in service, rather than in cure, a more volatile amine might be generated to expel it from the rubber.

It is the second of these ways of achieving control at source which offers the greatest potential for really effective fume control. Indeed, the approach is truly comprehensive: a pollutant which is destroyed or never created is not available for release either to the factory air or to other environments in the subsequent life of the product. The chemical interception of acrylonitrile during vulcanization may offer a practical basis for pollution control, whereas there are indications of accumulating experience in compound development to suppress the formation of, for example, nitrosamines, isothiocyanates or carbon disulphide. This is an area of increasing interest which will rely heavily on interactions between the researcher, compounder and raw materials supplier for its fullest development and exploitation.

6. CONCLUSIONS

Attention to the basic science of health and safety issues not only provides an insight into the current understanding of a complex and broad-ranging subject, but also helps to eradicate misconceptions. For example, seeing is not believing where air pollution is concerned, nor is exhaust extraction a comprehensive solution to pollution control. Dust does not cease to be a problem when it settles on the floor and fully vulcanized rubber is not necessarily free of toxic hazard. The case where settled dust reduces pedestrian friction can be contrasted with that where mechanical handling creates air pollution. Without doubt the interactions of this science are complex and some procedures designed to solve one problem may merely serve to introduce another.

The subject of health and safety is interdisciplinary and poses numerous challenges. Whilst many specializations may interact to provide solutions, the background and experience of the rubber technologist can provide the fulcrum of effective action. Many of the hazards in the rubber industry have their roots in the rubber and its unique characteristics such as its tack, friction, susceptibility to chemical change and its reluctance to wet and accept additives without high energy inputs. Those who work with rubber not only experience its problems but are in a unique position to contribute to the solutions required.

REFERENCES

1. Lawson, G., in *Developments in Rubber and Rubber Composites—2*, ed. C. W. Evans, Applied Science Publishers, London, 1983, Chapter 4.
2. Health and Safety Executive, *Occupational Exposure Limits 1985*, Guidance Note EH 40/85, HMSO, London.
3. BRMA Health Research Unit, *Toxicity and Safe Handling of Rubber Chemicals 1985*, 2nd (revised) edn, British Rubber Manufacturers' Association Ltd, Birmingham.
4. Nutt, A. R., *Toxic Hazards of Rubber Chemicals*, Elsevier Applied Science Publishers, London, 1984.
5. Thain, W., *Monitoring Toxic Gases in the Atmosphere for Hygiene and Pollution Control*, Pergamon Press, Oxford, 1980.
6. Anon, *Studies of Accidents in the Rubber Industry*, National Joint Council for the Rubber Manufacturing Industry, Manchester, 1952.
7. *BRMA Statistics*, British Rubber Manufacturers' Association Ltd, London.
8. Health and Safety Executive, *Rubber Health and Safety 1976–80*, HMSO, London, 1981.
9. Veys, C. A., in *Current Approaches to Occupational Health*, Vol. 2, ed. A. Ward-Gardener, Wright and Sons, Bristol, 1982.
10. Jacovone, M. T., *International Meeting on Health in the Rubber Industry*, BRMA, Edinburgh, June 1981.
11. Silant'ev, V. P., *Kauch. i Rezina*, (11), 1978, 53; *Int. Polymer. Sci. Technol.*, **6**(5), 1979, T/104.
12. Zegel'man, V. B., Karmatskii, Yu. L., Mikryukov, B. N. and Mit'kin, V. P., *Prov. Shin. RTI: ATI*, (7), 1979, 21; *Int. Polym. Sci. Technol.*, **7**(1), 1979, T/19.
13. Manning, D. P., *Ergonomics*, **26**, 1983, 3.
14. Barrett, G. F. C., *Rubber J.*, **131**, 1956, 685.
15. Brungraber, R. J., *An Overview of Floor Slip-Resistance Research with Annotated Bibliography*, NBS Technical Note 895, National Bureau of Standards, Washington, DC, Jan. 1976.
16. James, D. I., *Rubber Chem. Technol.*, **53**, 1980, 512.
17. Grieve, D. W., *Ergonomics*, **26**, 1983, 61.
18. James, D. I., *Ergonomics*, **26**, 1983, 83.
19. Brough, R., Malkin, F. and Harrison, R., *J. Phys. D: Appl. Phys.*, **12**, 1979, 517.
20. Tabor, D., in 'Advances in Polymer Friction and Wear', ed. L. H. Lee, *Polymer Science and Technology*, Vol. 5A, Plenum, New York, 1974, p. 5.
21. Roberts, A. D., *Prog. Rubber Technol.*, **41**, 1978, 121.
22. Spivey, A. M., 'Reduction of dust in working atmospheres by the use of improved product forms of rubber chemicals', PRI Conference, *Health and Safety in the Plastics and Rubber Industries*, University of Warwick, September/October 1980.
23. Parkes, H. G., Veys, C. A., Waterhouse, J. A. H. and Peters, A., *Brit. J. Indust. Med.*, **39**, 1982, 209.

24. PARKES, H. G., WHITTAKER, B. and WILLOUGHBY, B. G., *The Monitoring of the Atmospheric Environment in UK Tyre Manufacturing Work Areas*, British Rubber Manufacturers' Association Ltd, Birmingham, 1975.
25. HARRIS, R. L., ARP, E. A., SYMONS, M. J., VAN ERT, M. D. and WILLIAMS, T. M., 'Worker exposures to chemical agents in the manufacture of rubber tires and tubes', *112th Mtg Rubber Div. ACS*, Cleveland, OH, October 1977, Paper 40.
26. DORMAN, R. G. (Ed.), *Dust Control and Air Cleaning*, Pergamon Press, Oxford, 1974, Chapter 2.
27. LAPPLE, C. E., in *McGraw-Hill Encyclopedia of Science and Technology*, Vol. 9, McGraw-Hill Book Co., New York, 1971, pp. 661, 662.
28. LAPPLE, C. E. and SHEPERD, C. B., *Ind. Eng. Chem.*, **32**, 1940, 605.
29. HOLLAND, F. A., *Fluid Flow for Chemical Engineers*, Edward Arnold, London, 1973.
30. DORMAN, R. G., *Chem. Ind. (London)*, 1967, 1946.
31. DORMAN, R. G. (Ed.), *Dust Control and Air Cleaning*. Pergamon Press, Oxford, 1974, Chapters 3 and 9.
32. DINMAN, B. D., *Patty's Industrial Hygiene and Toxicology*, Vol. 1, eds G. D. Clayton and F. E. Clayton, Wiley, New York, 1978.
33. WALTON, W. H., in *Dust Control and Air Cleaning*, ed. R. G. Dorman, Pergamon Press, Oxford, 1974, Chapter 1.
34. VEYS, C. A., personal communication.
35. JACKMAN, P. J., in *Air Pollution Monitoring*, Proceedings of a Rapra Seminar, Rapra, Shawbury, 1978.
36. *Industrial Ventilation. A Manual of Recommended Practice*, 16th edn, American Conference of Governmental Industrial Hygienists, Lansing, Michigan, 1980.
37. WORWOOD, J. A., 'The engineering of dust and fume control systems', PRI Conference, *Health and Safety in the Plastics and Rubber Industries*, University of Warwick, September/October 1980.
38. DORMAN, R. G. (Ed.), *Dust Control and Air Cleaning*, Pergamon Press, Oxford, 1974, Chapter 4.
39. FLYNN, M. R. and ELLENBECKER, M. J., *Am. Ind. Hyg. Assoc. J.*, **46**, 1985, 318.
40. *Ventilation Handbook for the Rubber and Plastics Industries*, Rapra, Shawbury, 1970.
41. SHOTWELL, H. P., *Am. Ind. Hyg. Assoc. J.*, **45**, 1984, 749.
42. HUEBENER, D. J. and HUGHES, R. T., *Am. Ind. Hyg. Assoc. J.*, **46**, 1985, 262.
43. SOCHA, G. E., *Am. Ind. Hyg. Assoc. J.*, **40**, 1979, 1.
44. GOODFELLOW, H. D. and SMITH, J. W., *Am. Ind. Hyg. Assoc. J.*, **43**, 1982, 175.
45. *Granulated Chemicals for Modern Compounding*, Technical Report No. 50E, Rhein-Chemie Rheinau GmbH, Mannheim, FRG, 1974.
46. KASTEIN, B., *Rubber World*, **187**(3), 1982, 15.
47. GARNETT, A. A., *Plastics Rubber Int.*, **10**(5), 1985, 42.

48. KUCZKOWSKI, J. A. and GILLICK, J. G., *Rubber Chem. Technol.*, **57**, 1984, 621.
49. KEMPERMANN, Th., *Kautsch. u. Gummi Kunst.*, **31**, 1978, 234.
50. WILLOUGHBY, B. G., 'Reformulation or process control as a concept for the reduction of rubber curing fume', PRI Conference, *Health and Safety in the Plastics and Rubber Industries*, University of Warwick, September/October 1980.
51. LAWSON, G. and NEWELL, W., *Polymer Process Fumes—A Rationalised Approach towards an Understanding and Control of this Pollutant*, Rapra Members' Report No. 64, 1981.
52. ADAM, N. K., *Physical Chemistry*, Oxford University Press, London, 1958.
53. POPENDORF, W., *Am. Ind. Hyg. Assoc. J.*, **45**, 1984, 719.
54. KITTEL, C. and KROEMER, H., *Thermal Physics*, 2nd edn, W. H. Freeman and Co., San Francisco, 1980.
55. ARTEMOR, V. M., LITVINOVA, T. V., GULIL-OGLY, F. A. and ANDREEV, Yu. V., *Kauch. i Rezina*, (11), 1976, 16; *Int. Polym. Sci. Tech.*, **4**(4), 1977, T/4.
56. LUSTON, J., in *Developments in Polymer Stabilisation—2*, ed. G. Scott, Applied Science Publishers, London, 1980, Chapter 5.
57. GROTHER, E. W. and SMITH, C. H., 'Methods for the measurement of volatiles in polymers', *SPE 41st Annual Technical Conference, ANTEC 83*, Chicago, May 1983.
58. BERG, H., OLSEN, H. and PEDERSEN, E., *Gas Formation in the Vulcanising of Rubber*, Arbejdsmiljofondet, Copenhagen, 1982.
59. BERG, H., OLSEN, H. and PEDERSEN, E., *Gas Formation in the Vulcanising of Rubber II*, Arbejdsmiljofondet, Copenhagen, 1984.
60. LAWSON, G. and WILLOUGHBY, B. G., 'Atomspheric monitoring techniques applicable to the rubber industry', *116th Mtg Rubber Div. ACS*, Cleveland, OH, October 1979, Paper 29; *J. Soc. Env. Eng.*, **20**, 1981, 9.
61. WHITE, L. D., TAYLOR, D. G., MAUER, P. A. and KUPEL, R. E., *Am. Ind. Hyg. Assoc. J.*, **31**, 1970, 225.
62. RAPPAPORT, S. M. and FRASER, D. A., *Am. Ind. Hyg. Assoc. J.*, **38**, 1977, 205.
63. COCHEO, V., BELLOMO, M. L. and BOMBI, G. G., *Am. Ind. Hyg. Assoc. J.*, **44**, 1983, 521.
64. BAILEY, A. and HOLLINGDALE-SMITH, *Ann. Occup. Hyg.*, **20**, 1977, 345.
65. ROSE, V. E. and PERKINS, J. L., *Am. Ind. Hyg. Assoc. J.*, **43**, 1982, 605.
66. UNDERHILL, D. W., *Am. Ind. Hyg. Assoc. J.*, **45**, 1984, 306.
67. GOLLER, J. W., *Am. Ind. Hyg. Assoc. J.*, **46**, 1985, 170.
68. KRING, E. V., ANSUL, G. R., HENRY, T. J., MORELLO, J. A., DIXON, S. W., VASTA, J. F. and HEMINGWAY, R. E., *Am. Ind. Hyg. Assoc. J.*, **45**, 1984, 250.
69. WILLOUGHBY, B. G., *Europ. Rubber J.*, **166**(3), 1984, 49.
70. PITTS, J. N., Jr, LLOYD, A. C. and SPRUNG, J. L., *Chem. Brit.*, **11**, 1975, 247.
71. PATRICK, C. R., in *Preparation, Properties and Industrial Applications of*

Organofluorine Compounds, ed. R. E. Banks, Ellis Horwood, Chichester, 1982, Chapter 6.

72. DIVISION OF MEDICAL SCIENCES NATIONAL RESEARCH COUNCIL, *Particulate Polycyclic Organic Matter*, National Academy of Sciences, Washington, DC, 1972.

73. FRANKS, J., *Chem. Brit.*, **19**, 1983, 504.

74. HOWELLS, G. D. and KALLEND, A. S., *Chem. Brit.*, **20**, 1984, 407.

75. WILLOUGHBY, B. G., *International Meeting on Occupational Health in the Rubber Industry*, BRMA, Stratford-upon-Avon, May 1975.

76. NUTT, A. R., 'Measurement of some potentially hazardous materials in the atmosphere of rubber factories', WHO-NIEHS Conference, *Potential Environmental Health Hazards from Technical Developments in Rubber and Plastics Industries*, Research Triangle Park, NC, March 1976.

77. BUTLER, J. D., *Chem. Brit.*, **8**, 1972, 258.

78. *PTX Exhaust Gas Purifiers*, Engelhard Industries Ltd, Newport, Gwent.

79. WILLOUGHBY, B. G., *Prog. Rubber Technol.*, **46**, 1984, 143.

80. *Toxicity Information Related to the Handling and Processing of 'Vamac'*, Du Pont Co., Wilmington, Delaware, Feb. 1983.

81. ASHNESS, K. G., LAWSON, G. and WILLOUGHBY, B. G., 'Origin and reduction of toxic vapours from rubber vulcanisation', *Rubbercon '81* Harrogate, June 1981.

82. ASINGER, F., *Mono-olefins Chemistry and Technology*, translator B. J. Hazzard, Pergamon Press, Oxford, 1968.

83. BRIGHTON, C. A., PRITCHARD, G. and SKINNER, G. A., *Styrene Polymers: Technology and Environmental Aspects*, Applied Science Publishers, London, 1979.

84. CROMPTON, T. R. and MYERS, L. W., *Plastics and Polymers*, **36**, 1968, 205.

85. ASHNESS, K. G., LAWSON, G., WETTON, R. E. and WILLOUGBY, B. G., *Plast. Rubb. Process. Appln*, **4**, 1984, 69.

86. WILLOUGHBY, B. G. and LAWSON, G., 'Laboratory vulcanisation as an aid to factory air analysis', *116th Mtg Rubber Div. ACS*, Cleveland, OH, October 1979, Paper 28; *Rubber Chem. Technol.*, **54**, 1981, 311.

87. WILLOUGHBY, B. G., *Rubber World*, **187**(3), 1982, 26.

88. RAPPAPORT, S. M. and FRASER, D. A., *Anal. Chem.*, **48**, 1976, 476.

89. PELOSI, L. F., MORAN, A. L., BURROUGHS, A. E. and PUGH, T. L., *Rubber Chem. Technol.*, **49**, 1976, 367.

90. HILTON, A. S. and ALTENAU, A. G., 'Mass spectrometric identification of 2-mercaptobenzthiazole sulphenamide accelerators in rubber vulcanisates', *103rd Mtg Rubber Div.*, *ACS*, Detroit, May 1973, Paper 22.

91. CAMPBELL, R. H. and WISE, R. W., *Rubber Chem. Technol.*, **37**, 1964, 635.

92. CAMPBELL, R. H. and WISE R. W., *Rubber Chem. Technol.*, **37**, 1964, 630.

93. CORAN, A. Y., *Rubber Chem. Technol.*, **37**, 1964, 679.

94. CRAIG, D., DAVIDSON, W. L. and JUVE, A. E., *J. Polym. Sci.*, **6**, 1951, 177.

95. COLEMAN, M. M., SHELTON, J. R. and KOENIG, J. L., *Rubber Chem. Technol.*, **46**, 1973, 957.
96. WILLOUGHBY, B. G., 'Investigation of additive performance and perspective for process control', *8th Scandinavian Rubber Conference SRC8S*, Copenhagen, June 1985.
97. MOORE, C. G., MULLINS, L. and SWIFT, P. McL., *J. Appl. Polym. Sci.*, **5**, 1961, 293.
98. KOCHANOVA, O. M., BLOKH, G. A., KOKMAN, F. S., STRELOK, I. M., LEVINA, S. A., ERMOLANKA E. F. and MALASHEVICH, L. N., *Soviet Rubber Technol.*, **29**, 1970, 18.
99. LOAN, L. D., *Rubber Chem. Technol.*, **49**, 1967, 149.
100. MOORE, C. G. and WATSON, W. F., *J. Polym. Sci.*, **19**, 1956, 237.
101. WERWERKA, D., HUMMEL, K. and INSELBACHER, W., *Rubber Chem. Technol.*, **49**, 1976, 1142.
102. APOTHEKER, D., FINLAY, J. B., KRUSIC, P. J. and LOGOTHETIS, A. L., *Rubber Chem. Technol.*, **55**, 1982, 1004.
103. DELABY, R., DAMIENS, R. and SEYDEN-PENNE, R., *Compt. Rend.*, **238**, 1954, 121.
104. BLAKE, E. S., *J. Am. Chem. Soc.*, **65**, 1943, 1267.
105. SMITH, P. A. S., *Open-chain Organic Nitrogen Compounds*, Vol. 1, W. A. Benjamin Inc., New York, 1965, Chapter 2.
106. *The Rubber Industry*, Vol. 28 of the IARC Monographs on the Evaluation of the Carcinogenic Risk of Chemicals to Humans, International Agency for Research on Cancer, Lyons, 1982, pp. 121–47.
107. FAJEN, J. M., CARSON, G. A., ROUNDBEHLER, D. P., FAN, T. Y., VITA, R., GOFF, V. E., WOLF, M. H., EDWARDS, G. S., FINE, D. H., REINHOLD, V. and BIEMAN, K., *Science*, **205**, 1979, 1262.
108. YAEGER, F. W., VAN GULICK, N. N. and LASOSKI, B. A., *Am. Ind. Hyg. Assoc. J.*, **41**, 1980, 148.
109. IRELAND, C. B., HYTREK, F. P. and LASASKI, B. A., *Am. Ind. Hyg. Assoc. J.*, **41**, 1980, 895.
110. EMBER, L. R., *Chem. Eng. News*, **58**(13), 1980, 20.
111. GROVES, J. S. and SMAIL, J. M., *Brit. J. Ophthal.*, **53**, 1969, 683.
112. JEFFERSON, A., SMITH, R. W. B. and WILLOUGHBY, B. G., *Chem. Ind. (London)*, 1986, 244.
113. SMITH, S., in *Preparation, Properties and Industrial Applications of Organofluorine Compounds*, ed. R. E. Banks, Ellis Horwood Ltd, Chichester, 1982, Chapter 8.
114. ITO, H., *Japan Plastics Age, 7*, 1969, 25.
115. *The Genitron Range of Blowing Agents*, Fisons Industrial Chemicals, Cambridge.
116. NUTT, A. R., *Prog. Rubber Technol.*, **42**, 1979, 141.
117. SCHMIEGEL, W. W., *Angew Makromol. Chem.*, **76/77**, 1979, 39.
118. PARISER, R., *Kunststoffe*, **50**, 1960, 623.
119. TRUCE, W. E., KLINGER, T. C. and BRAND, W. W., in *Organic Chemistry of Sulphur*, ed. S. Oae, Plenum Press, New York, 1977, Chapter 10.

120. Buus, H., 'Toxicological evaluation of chemical substances generated by sulphur vulcanisation of rubber', *8th Scandanavian Rubber Conference SRC85,* Copenhagen, June 1985.
121. Buus, H., *Rubber Vulcanisation and Health Hazards,* Arbejdsmiljofondet, Copenhagen, 1986.
122. Munn, A., *Rubber Industry,* **8,** 1974, 19.

INDEX

Abrasion resistance, 36
Accidents, 256–62
  falling, 260
  frequency of, 257–9
  muscular stress, 259
  records, 258
Acrylonitrile, 295
Aktiplast, 142
Alcohol/gasoline blends, 184
Alkyl isothiocyanates, 295
Aniline point, 124
Anionic polymerization, 1–56
API test, 170–4
Aqueous acid conditions, 183
Artificial lift, 162
ASTM Classification System, 164, 187
Automotive industry, seal performance requirements, 187–8

Ba/Mg/Al catalyst, 20, 22, 23, 25, 26, 29, 31
Banbury mixer, 198, 200, 203, 233
Batch mixer discharging, 208–10
Batch mixing, 198–210
  uniformity variation, 210
Benzo(a)pyrene, 285
Benzyltriphenylphosphonium chloride (BTPPC), 297
4,4'-Bis(diethylamino)benzophenone, 16
1,3-Bis(phenylethenyl)benzene, 15
Biton GF, 182
Black incorporation time (BIT), 242
Blow out preventers (BOPs), 160

Bore holes, 159
Branching, 12–13
Butadiene–acrylonitrile rubbers, 59
Butadiene–styrene copolymerization, 25–7
Butadiene–styrene copolymers, variation of composition with conversion, 25–6

Capillary rheometers, 241
Carbon black dispersion, 207, 243
Carboxylated nitrile rubber (XNBR). See Nitrile rubber (NBR)
Chain branching, 13
Chain end functionalization, 16
Chain extension, 18
Chamber loading optimization, 231
Chlorofluorocarbons, 285
Choke manifold, 161
Christmas tree stack, 161
Coefficient of kinetic friction, 262, 263
Comonomer sequence distribution control, 7–9
Compaction process, 226
Compression set, 126
Computerized data handling, 285
Contaminant monitoring, 282–5
Continuous mixing, 215–20
  computer control, 219
  direct product production, 219–20
  machine utilization, 216
  machines, 218–20
Control at source
  dust hazards, 274–5
  vapours, 299–300

Corporate Average Fuel Economy
(CAFE), 33, 38
Corrosion inhibitors, 174, 189
Coupling agents, 15
Crystalline melting temperature, 4
Curing system, vapours, 290–6

Dark-field reflected light (DFRL)
microscopy, 243
Decomposition products, 296–8
Delta Mooney, 240
Dialkylphenylenediamine, 297
Dienes
blends of solution polymer with,
43–5
organolithium initiation of, 2–3
polymerizations, 2
Diesel fuel, 186
Dipiperidinoethane (DIPIP), 3, 6
Dispergum 24 (DOG), 142
Dispersion process, 226–8
Divinylbenzene (DVB), 12–13
Drilling rig, 160
Dump criteria, 229–30
Dump extruders, 211–13
Dunlop resilience, 126
Dust generating potential, 274
Dust hazards, 264–75
control at source, 274–5
Dust suppression, 274
Dutrex 729, 143

E-SBR, 8, 27, 28, 40, 43, 44
Elastomers, solubility parameters,
152
Electrical storms, 194
Emulsion SBR, 1, 33
Energy inputs, 202–3
Energy requirements for viscosity
reduction, 148
ENR-25, 89–91, 100, 101, 103,
104–6, 111, 112
ENR-50, 89–91, 94, 95, 97, 99–101,
103–6, 111, 114
Epichlorhydrin rubber (ECO), 166,
167, 179

Epoxidized natural rubber (ENR),
87–117
air ageing, 109–111
air permeability, 104
applications, 111–15
black filled vulcanizates, 100–3
black mixing, 93
bonding, 105–8
chemistry of epoxidation, 88
compounding, 93–4
cured adhesion, 106
damping properties, 114
dynamic properties, 115
epoxide analysis, 89–90
epoxide distribution, 90–1
general applications, 115
gum vulcanizates, 100
hysteresis, 104
mastication, 93
mix cycle, 93–4
oil resistance, 103–4
ozone resistance, 111
peroxide vulcanization, 97–8
physical properties, 98–111
polymer manufacture, 88–9
polymer structure, 89–91
production, 88–9
raw properties, 91–3
rebound resilience, 106
silica filled, 104–5
steel bonding, 108
stereochemistry, 90
strain crystallization, 98–100
strength properties, 100–3
sulphur formulations, 95–7
tack, 106
vulcanization, 94–8
wet grip properties, 109
2-Ethylhexyl acrylate, 82
Evolutionary operation (EVOP), 236
Explosive decompression effect, 169,
189
Explosive decompression resistance,
169–70
Extrusion improvement using fatty-
acid soaps, 151
Extrusion resistance test, 189
Extrusion resistance under high
pressure, 170–4

Factice, 153
  areas of use, 154
  effects on vulcanized properties,
    154–5
  grades available, 153–4
Falls, injuries caused by, 260
Fatty-acid soaps, 142–6
  blends with chemical peptizers,
    146–9
  characteristics of, 144
  esters of, 151–2
  internal lubrication, 144–5
  materials examined, 142–3
  processability, 143
Feeding problems, 207–8
Feeding techniques, 204–5
FEP, 164
Fina Para, 120
FKM fluoroelastomer, 170
Flash point, 124
Fluorocarbon rubbers, 167
Fluoroelastomers, comparison of
  NBR with, 175–7
Fluorosilicone rubber, 166
Fox equation, 36
Friction, classical laws of, 260
Frictional force, 260
Functional terminated polymers,
  15–16

Gas chromatography, 285, 287
Gas production, 159–64
Gas transfer mould (GTM), 283, 287
Glass transition temperature, 4,
  34–6, 44, 91–3
Grafted polymers, 81–2
Granulation, 215–16
Green strength enhancement, 145

Hancock's 'Pickle', 120
Hardness, 125, 129
Health and safety, 253–306
  see also Accidents
Heat exchange, 228–9
Heat transfer, 202
Hevea brasiliensis, 120

High-trans SBR (HTSBR). See
  HTSBR
High-vinyl BR (HVBR), 6
  nature of chain ends, 22
Highly saturated NBR (HSNBR),
  59–63
Homogenizing resins, 152–3
HTSBR, 17–33, 46
  absorption patterns, 28
  characteristics of, 47
  crystalline melt temperature, 28–9
  crystallinity in, 27–33
  crystallization rate, 32
  cycles to failure versus strain
    amplitude, 51
  green strength, 50
  melting temperature, 30
  molecular weight, 19
  molecular weight distribution,
    18–19
  percentage crystallinity, 32
  physical properties, 48
  strain-induced crystallization
    behaviour, 29, 31
  stress–strain curves, 49
  styrene sequence distribution
    characterization, 27
  synthesis of, 17–20
  tread properties of blends, 48
  tread vulcanizate properties, 47
  uncrosslinked, 27–9
Hydrogen sulphide, 180–3, 189
Hydrogenated nitrile (HNBR), 178

Incorporation process, 224–6
  mechanisms of, 225
Inertia capture of airborne
  particulates, 269–70
Instrumented portable skid tester
  (IPST), 44
Integral rubbers, 40
ISAF(N220), 103

Jack up rigs, 160

K2A intermix conditions, 136
Kalrez®, 167, 183

Lamp formation in batch mixing, 206–7
Lewis bases, 3
Living polymers, 20–4
Local exhaust ventilation (LEV), 270–4
Low temperature service, 167–8
Low vinyl solution SBR (LVSBR), 9, 42

Mass spectrometry, 285, 287
Mastication, 133
Medium-vinyl butadiene rubber (MVBR), 40
Medium-vinyl SBR (MVSBR), 11, 45–6
Methanol resistance, 183–4, 189
Microprocessors, 235
Mill room equipment, 193–220
Miniature internal mixers, 242
Mists, 275–6
Mixing energy, 147
Mixing equipment, 232–6
  rotor design, 232–5
Mixing process, 221–51
  application of science, 222
  development needs, 221–2
  dispersive, 224
  distributive, 223
  efficiency of, 232
  extensional flow, 225–6
  operating variables, 230–2
  order of ingredient addition, 231
  process control, 235–6, 246–7
  process variables, 228–30
  product testing, 243–4
  quality aspects, 244–7
  scale-up, 231–2
  shear flow, 226
  study of, 228
  types of, 223–4
  see also Processability testing
Molecular structure, 238–9
Molecular weight distribution (MWD), 11–13, 43, 239
  bimodal, 15, 46
  broadening, 12–13, 18

Molecular weight distribution (MWD)—contd.
  monodisperse, 15
  narrow, 18
Monitoring
  atmospheric, 284
  contaminant, 282–5
Monsanto Processability Tester (MPT), 241
Mooney scorch time, 95
Mooney test, 239–41, 247
Mooney torque peak (PMT), 240
Mooney viscosity, 91, 124, 127, 147
Muscular injury, 259
MVX continuous mixer, 218–20

National Association of Corrosion Engineers (NACE), 175
NDPA, 295
Nitrile rubber (NBR), 57–85, 166
  acrylonitrile content, 79
  bound-antioxidant, 63–8
    applications, 68
    attaching the antioxidant, 64–5
    physical properties of, 66
    test results, 65–8
  carboxylated, 68–70, 166, 178
    applications, 70
    crosslinking, 68–9
    internally retarded, 70
    PVC blends, 76
    scorch control, 69–70
  characteristics of, 57–9
  classification, 78
  comparison with fluoroelastomers, 175–7
  future prospects, 82–3
  general properties of, 58
  grafted polymers, 81–2
  highly saturated, 59–63
  hydrogenated, 59–63
    physical properties, 62
    typical formula, 62
  liquid grades, 81
  miscellaneous developments, 80–2
  obtaining saturation, 60
  polymer modifications, 59
  powdered grades, 80–1

Nitrile rubber (NBR)—*contd.*
  PVC blends, 75–8
    mechanical preparation, 76
  PVC modification, 77–8
  rationalized grades, 79–80
  saturated, properties of, 60
  standard grades, 78–80
  thermoplastic, 71–5
    applications, 75
    commercially available, 71
    compression set, 75
    physical properties, 72
    two-phase structure, 71
    volume swell, 73
Nitrogen oxides, 286
N-Nitrosamines, 294
Nuclear magnetic resonance, 7, 10

OESBR 1712, 112
Oil drilling, 160
Oil exploration drilling, 159
Oil production, 159–64
Oil wells, 161–2
  additives, 174–80
  enhancement or stimulation, 162
Oilfield seals, 159–91
  applications range, 162
  compatibility requirements, 168
  elastomer selection, 164–8
  elastomer specification, 186–90
  specific problem areas, 168–86
Organolithium polymerization
    catalysts, 2
Oxidation process, 296–8

Particle dynamics, 264–5
Particle projection, 266–8
Particle size effects, 266
Particles in air, 264–8
Particles inhalation or ingestion, 270
Particulate deflection and capture,
    269–70
Pelletizer, 212
Pelletizer head, 213
Peptization, 133
  in batch, 137

Peptizers, 131–49
  evaluation, 136
  fatty-acid soap blends with, 146–9
  level of use, 132
  levels of, 133–4, 141
  processing trials, 132–40
  types in use, 131
  usefulness of, 140
Perfluoroelastomers (PFE), 167, 184
  resistance to amine-stabilized oil,
    179
Peroxide vulcanization, 97–8
Peroxidized diesel oil, 186
Petroleum oils, 122–31
Phenylenediamines, 296–7
Pipe lines, 162
Plasticizers, 122–31
  general characteristics, 122–3
  solubility parameters, 152
  summary of use of, 156
  terminology, 123–4
Polyacrylate rubber, 166
Polybutadiene, 1, 33
  trans-1,4-content, 21, 22
  effect of modifiers on vinyl content,
    5
  glass transition temperature and
    melting temperature, 4
  glass transition versus percentage
    vinyl content, 7
  microstructure variations, 3–7
  microstructure versus mole ratio of
    Ba/Mg, 20
  reactive anionic end groups of, 21
Trans-1,4-Polybutadiene, 29, 33
Poly(butadiene-co-styrene), 1
Polycyclic aromatic hydrocarbons,
    285
Polymer blends, 197–8
Polymer characterization, 2–16
Polymer synthesis, 2–16
Polyphenylene sulphide, 180
Polystyrene, gel permeation
    chromatograms, 23
Polystyrene–polybutadiene diblock
    copolymer, 23
Post-mixer, shape and cool, 210–13
Powder dispersion, 207

Preblending, 216
Precision roller die, 212
Preconditioning, 195
Preheating, 195, 198–200
Premixers, 217–18
Prevulcanization inhibitors (PVI), 95
Process aids, 119–22
    mechanisms of, 122
    other than viscosity reducers,
        149–56
    parameters for, 120
    summary of use of, 156
    uses, 120
    see also Factice; Viscosity reducers
Process control, mixing process,
        235–6, 246–7
Processability testing, 236–44
    requirement of, 237–8
Production rig, 160
Production tubing, 161
Production uniformity, 194
Proton NMR, 10, 27, 28
PTFE, 164
PVC, 263
    NBR as modifier for, 77–8
    see also Nitrile rubbers (NBR)

Quality aspects, mixing process,
        244–7

Ram pressure, 200–2, 230
Raw materials
    quality requirements, 245–6
    testing, 238–9
Renacit VII, 143
Reversion resistance, 146
Rheological properties, 238–9
Rheometer data, 128
Rotor speed, 202–4, 230
Ryton, 167

Safety issues. See Accidents; Health
        and safety

Saybolt viscosity, 125
Sealing applications. See Oilfield seals
Semi-submersible rigs, 160
Shoe soling, 262
Slabber heads, 211
Slabber/strainer head, 211
Slipping falls, prevention of, 260
SMR 10, 133, 147
SMR 20, 137, 138
SMR CV, 131
SMR GP, 131
SMR LV, 131
Sodium nitrite, 295
Solution SBR, 1
    anionic polymerization, 9–11
    prospects for, 51–2
    random, 8, 25, 41
    sequence length of comonomer
        units, 7
    Sn-coupled, 14
    structural control, 3
    styrene sequence distribution, 9–11
    technological properties of, 33–51
Steam resistance, 189
    of elastomers, 183
Stock preparation, 194–8
Stokes equation, 266
Strainer slabber, 212
Stress relaxation testing, 242
Struktol A60, 142, 143
Struktol A82, 143
Styrene–butadiene rubbers (SBR)
    hydrocarbons found in, 291
    in organic solvents. See Solution
        SBR
    macrostructure variations, 11–14
    medium-vinyl (MVSBR), 11, 45–6
    molecular features of, 39
    proton NMR spectra, 10
    random solution, 9
    reactivity ratios, 26–7
    Sn-coupled solution, 14
    structural control, 46
    styrene incorporated as long
        sequences in, 11
    vinyl-substituted, 42
    see also Solution SBR
Sulphur formulations, 95–7

Superior Processing (SP) natural rubber, 155–6
  comparative viscosities and extrusion behaviour, 156
Synthetic rubbers, development of, 121

Teflon, 164, 167
Temperature measurement, 205–6
Tensile strength, 125
Tetrahydrofuran (THF), 6
Tetramethylethylenediamine (TMEDA), 6, 11, 36
Tetramethylsuccinonitrile, 296
TFE/propylene rubber, 167, 168, 178–81, 184
Thermoplastic rubbers, 198
Tin-coupled polymers, 14
TMS (Turner, Moore and Smith) rheometer, 240
Torque rheometers, 242
Two-roll mills, 214–15
Two-roll sheeter, 212
Tyre carcass properties, crystallizable solution rubbers, 49–51
Tyre tread formulations, 33, 111–14
  abrasion resistance, rolling resistance and wet traction, 45
  conventional tread rubber behaviour, 38–9
  influence of styrene and vinyl contents, 40–3
  performance and properties relationship, 35
  performance properties, 44
  rolling resistance and wet traction and wear, 34–43, 112
  solution polymers for, 34
  solution rubbers developed for improved properties, 45
  structural features influencing, 34–9

Vapour Hazard Index (VHI), 277
Vapours
  areas of particular concern, 294–5
  background air, 285
  competing reactions, 292–5

Vapours—contd.
  contaminant measuring, 282–5
  control at source, 299–300
  curing system, 290–6
  generation and dispersal, 276–80
  generation from liquid, 277
  generation from rubber, 278–80
  generation of new products, 292–3
  hazards, 275–85
  measures of concentration, 280–2
  mixtures and their analysis, 280–5
  suppression by reaction, 295–6
  toxic effects, 298–9
  velocity, 276–7
  workplace, 285–300
  see also Vulcanization fume
Variable speed mixing, 203
Velocity profiles, 271
Ventilation effects on airborne contamination, 270–1
Vinyl-BR rubbers, properties of, 40
Vinylcyclopentane formation mechanism, 6
Vinyl-substituted SBR, 42
Viscosity gravity constant (VGC), 123
Viscosity measurements, 138
Viscosity reducers, 131–149
  energy requirements for, 148
  results achieved, 137
  role of, 131
Vulcanizate properties, 124, 139, 147
Vulcanization fume, 287–90
  contributions from base polymer, 287–9
  dimers, 287–9
  low molecular weight impurities, 287
  residual components, 289–90
  trimers, 287–9
  wide vapour range, 290

Weighing accuracy, 196–7
Wetting process, 224
Workplace safety, rubber sole in, 262

XNBR gum compounds, 69

Zinc stearate, 142

anterior pituitary (AP) function,
  316, 318-20
cardiovascular changes, 321
circadian rhythms, 306
corticotrophic development, 313
fetal adrenal cortex, 311-13
growth hormone (GH),
  306, 310-11
luteinizing hormone (LH), 306
prolactin (PRL), 306
scrotal closure (test), 308-9
seasonal breeding, 304-6
thyroid, 309-11

Stress response, 294-6, 314-17
The affects (SDN), 288
Suckling, 306
Synchronization of labour, 285

Temperature, regulation, 293-4
Testosterone (T), 308
The pineal gland, 306
Thyroid, stimulating hormone
  (TSH), 309-11
Thyroxine (T4), 309-11
Thyrotrophin releasing hormone
  (TRH), 309-11

Ultrasound, 282-5, 291, 296, 301-2
Urinary oestriol, 266-7
Uterine contractility, 285-6, 289

Vagal tone, 287
Vasopressin, arginine (AVP),
  314-15, 317

Weight, at birth, 266

X-zone, fetal adrenal, 311